W9-CAL-423

ATOM LAND

Also by Jon Butterworth

Most Wanted Particle

ATOM LAND

A Guided Tour Through the Strange
(and Impossibly Small)
World of Particle Physics

Jon Butterworth

NEW YORK

To Ann and Keith

Atom Land: *A Guided Tour Through the Strange (and Impossibly Small) World of Particle Physics*

Originally published in the UK as *A Map of the Invisible* by Jon Butterworth © 2017.
First published in North America by The Experiment, LLC, in 2018.

The Experiment, LLC | 220 East 23rd Street, Suite 600 | New York, NY 10010-4658
theexperimentpublishing.com

Library of Congress Cataloging-in-Publication Data

Names: Butterworth, Jon, author.
Title: Atom land : a guided tour through the strange (and impossibly small) world of particle physics / Jon Butterworth.
Description: New York, NY : The Experiment, LLC, [2018] | "Originally published in the UK as A Map of the Invisible by Jon Butterworth in 2017"--Title page verso.
Identifiers: LCCN 2017052472 (print) | LCCN 2018004229 (ebook) | ISBN 9781615193745 (ebook) | ISBN 9781615193738 (cloth)
Subjects: LCSH: Particles (Nuclear physics)--Popular works.
Classification: LCC QC793.26 (ebook) | LCC QC793.26 .B747 2018 (print) | DDC 539.7/2--dc23
LC record available at https://lccn.loc.gov/2017052472

ISBN 978-1-61519-373-8
Ebook ISBN 978-1-61519-374-5

Cover design by Sarah Smith
Text design in India by Integra Software Services Pvt. Ltd, Pondicherry
Author photograph by Paul Clarke

Manufactured in the United States of America

First printing March 2018
10 9 8 7 6 5 4 3 2 1

Contents

A Note on the Maps

The maps within this book should be seen as an aid to memory rather than a detailed representation of particle physics. There is a rough direction of increasing energy (and decreasing size) from west to east and increasing complexity from south to north, but there are gray areas and ambiguities. Sometimes the energy scale is binding energy, sometimes mass, and there are unavoidable inconsistencies even so. For example, the photon belongs on Bosonia in the east, even though its presence is felt far to the west. The tau and muon should arguably be farther east than the up, down, and strange, based on their masses, although one does really have to cross Lambda QCD to get to the quarks. Allegory and analogy can help understanding, but are misleading if pushed too far. Enjoy yourself, but tread with care.

PROLOGUE

The Journey Begins

Consider a thought experiment: Take an apple and cut it in half, then cut it in half again, and again, and keep doing that. What will you end up with?

Or, less destructively, peer more and more closely into an apple: What structure is revealed? Is everything made up of a small set of common constituents—call them elements, or atoms—arranged in various different ways? If so, what happens if I look more closely at those constituents? Are they made of something even smaller?

What is eventually revealed is a landscape. A landscape inhabited by strange and wonderful particles. In this landscape are islands of complexity linked by networks of communication, emerging from the oceans of our ignorance. The landscape starts from everyday life, with an apple, for example, and extends to the wild frontier of the almost impossibly small.

To get from an apple to the impossibly small, we will need to set sail. We will need a boat, a craft that will stand for the microscopes, particle accelerators, and other machinery that extend our vision beyond the capacity of the naked eye and into the heart of the atom and beyond. How far can we sail? Is there an end to this invisible world? Are there indivisible particles, of which everything else is made, or can we carry on forever, finding ever-smaller stuff, sailing ever farther eastward?

These questions have been discussed for millennia, and addressing them is one of the goals of physics. The answers, as far as we know them, are found in the strange, invisible landscape that we will explore and map in this book.

The Name of the Game

The science of the very smallest things is commonly called "particle physics." No name is perfect, and this one is confusing in a number of ways.

The word "particle" is potentially misleading. Physicists study particles of sand, pollutants, dust—in space and in the atmosphere—and other small clumps of stuff that have nothing directly to do with the ultimate constituents of matter.

Sometimes particle physics is called "elementary particle physics" to distinguish it from the study of these composite particles. But this isn't much help, since protons and neutrons—important, tiny particles that are crucial features in the landscape of the subject—aren't elementary. At some point we may even find out that the fundamental particles of our current theories aren't elementary, either, although studying them is definitely "particle physics." Elementary particle physics is also disfavored as a term for research groups or university courses because it makes the subject sound too easy. A student who enrolled in an Elementary Particle Physics course might get a shock when confronted with the equations relating to the particles we will be searching for.

"High-energy physics" is a commonly used alternative, and it is true that the direct approach to particle physics—basically, smashing things together in huge colliders to see what happens—involves a lot of energy. But some of the key experiments actually rely on hunting for very, very rare, very low-energy particles. Physicists hide their detectors deep underground to try to evade tiny amounts of noise, and every ultra-low-energy jitter is a cause for annoyance or excitement. Indirectly, these experiments do tell us something about what goes on at high energies, but calling them "high-energy physics" somehow seems a bit inappropriate.

Another problem with "high-energy physics" as a name is that nuclear physicists, astrophysicists, plasma physicists, and many more all deal with energies that are much higher than those used to probe the limits of particle physics. The energy of a collision in the Large Hadron Collider, which at the time of writing is the highest-energy particle collider ever built, is tiny compared to the energy that is released in a functioning nuclear reactor, which itself doesn't even register when compared to an exploding star.

Whatever the name we give the subject, the investigation is a practical one, liberated from the realms of speculative philosophy by experiments, beginning with the human eye, then the microscope, and continuing with powerful particle accelerators and other precision instruments. Each new generation of experiments opens up to human inquiry new landscapes of the very small, and lets us make maps to guide the way deeper into the heart of matter.

But in the end, the question is the same, then and today: What is the universe made of, really, when you get right down to it?

The Standard Model

The current answer is encapsulated in a theory going by the understated (frankly, dull) name of the "Standard Model," which summarizes the current state of our knowledge of the fundamental forces and constituents of matter—the area of science generally known as particle physics. This is a theory (it is really more of a theory than a model, although the terms mean different things to different people) that is the product of many decades of work and research, and it successfully describes a vast range of physical phenomena.

When I was growing up it was quite common for Indian or Pakistani restaurants to call themselves "Standard," as in Standard Tandoori or Standard Balti Palace. Given that I grew up in Manchester, England, home of the Curry Mile, the "standard" was extremely high, and was something to which a new restaurant would aspire. I think the Standard Model should be seen in similar terms. The modest name is a mark of quality. Any new theory that may come along has a high standard to meet.

An example of the power of the Standard Model is the success it achieved in 2012 with the discovery of the long-predicted Higgs boson. This is a kind of object new to us, unprecedented in nature, and it is essential to the mathematical consistency of the theory. But for now it is enough to say that the discovery of the Higgs boson was an amazing vindication of the ideas behind the theory. Remarkably, we now have a self-consistent theory in which the smallest objects are indeed infinitely small. This theory can describe phenomena over an enormous range of energies and distances, a range that was dramatically extended by the discovery of the Higgs boson.

The ideas behind the Standard Model are elegant and mathematical, and considering the enormous range of observations it can describe, it is remarkably compact and simple. Each individual idea going into it can be understood, in outline at least, by anyone. But there are several important and unfamiliar ideas, and building up the overall picture, the way it all fits together, is challenging.

The Standard Model is necessarily dynamic. It may need to be changed when new data comes along. That flexibility does not, however, alter the fact that it is an effective theory, beautifully describing a vast array of data. It contains truth—just not all the truth.

This book is a quest for the truth. Or a quest for as much truth as we know.

It will take you on eight—maybe eight-and-a-half—expeditions deep into the heart of the material universe. These expeditions will together reveal and explore the smallest constituents of matter, examine how they behave (they often behave quite strangely), and identify the forces that bind and break them. This is the story of our world, and indeed, our universe. It is from these building blocks that the molecules and materials of everyday life, and the stars and galaxies beyond, are formed.

As we explore new territory on the edge of physical knowledge, the lands we discover will be named and located in relation to each other. Quarks, bosons, hadrons, and the rest will be set down in what is in effect an illustrated glossary that also provides a framework in which the ideas of the Standard Model can be placed. The features we encounter may appear a little arbitrary sometimes, but they reflect the best evidence we have so far. And while—as far as we know—some of the features of the Standard Model are in fact arbitrary, others have deep explanations, and the whole structure has a high degree of elegance and economy, at least compared to any preceding theory.

Before embarking upon our journey, I should note that this is only one possible journey to the frontiers of science. It is a reductionist approach, and we know that it will not reveal the whole story, in the sense that, even if it were to reveal a so-called "theory of everything" (which may or may not fit on a T-shirt), it would still leave a lot unknown. Whatever tiny constituents are revealed by particle physics, we already know that their interactions—and the ways they behave in large numbers—reveal deep underlying principles and complex behaviors that aren't necessarily obvious from the so-called "fundamental" laws. There is new physics—not to mention chemistry, biology, and the rest—to be found here as well. However, knowledge of the structure of matter at the smallest accessible distances is important, and it is surely one of the most exciting frontiers of science, and that's where we are going. The maps we make will also reveal some of those startling and beautiful principles that seem to apply far and wide across nature.

And, like any early explorers' maps, there are edges. The Standard Model may be complete, but our understanding of physics is not. The final voyage will take us off into the unknown, wary of sea monsters and the distracting calls of the siren, in search of more answers.

EXPEDITION I

Sea Legs

A boat, and what it's made of – Gulls, dolphins, and interference – A lesson from the pilot – Some skeptical crew, impatient to sail – The pilot goes on and on, with lasers – The crew are convinced – Not your usual kind of field – Short distances, high energies, and the relationship between them – The importance of choosing your path

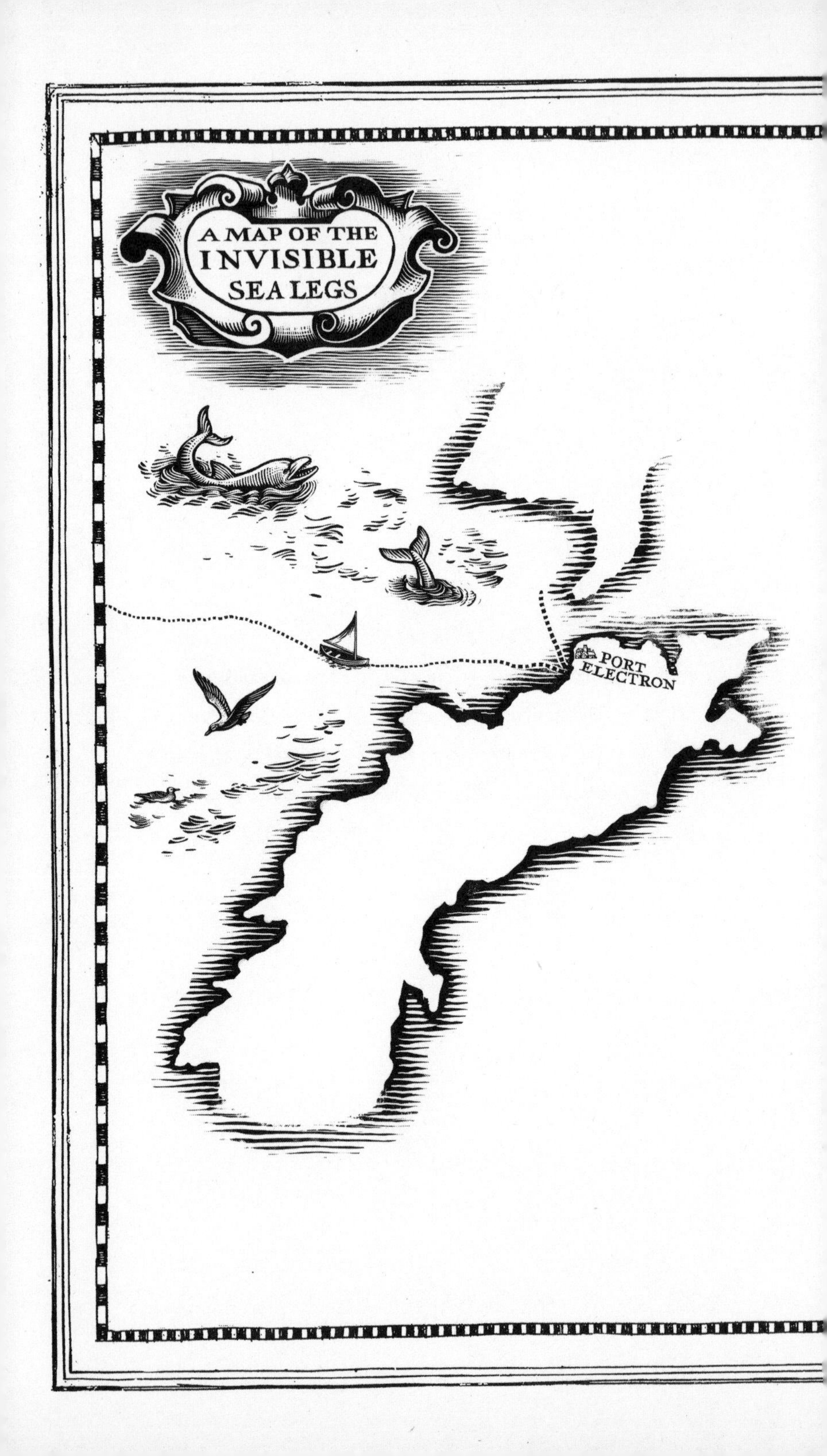
A MAP OF THE INVISIBLE SEA LEGS
PORT ELECTRON

N
W
E
S
LOW ENERGY
HIGH ENERGY

I

Setting Sail

Having obtained a small but seaworthy vessel, with a makeshift crew of professional physicists and curious amateurs, we set sail. The hold contains provisions, a mix of scientific equipment, and a guitar. We have some theories to test, and we need data. We hope, like Darwin on the voyage of the *Beagle*, to find that we will be improved by a journey to distant countries.

We enter the mysterious seascape of the invisibly small from the west. On our map, the western fringes are objects at a human scale. As we sail eastward we will shrink further, gazing from the bow of our boat into the heart of matter, mapping the otherwise invisible.

Most things are made of smaller things. Our boat is made of wood, metal, fiberglass. It doesn't take much effort to see that these materials are themselves made of smaller stuff: splinters of wood, glass fibers, plastic. The glass fiber strands, the thickness of cotton thread, are made of silica. Each one of these strands is made of silicon atoms bound to oxygen atoms, with two atoms of oxygen for each of silicon—silicon dioxide. A silicon atom is a billion times smaller than the thickness of the thread. If each silicon atom were the size of one of the peas in the ship's galley, the fibers would have a diameter close to that of the Earth.

Each atom consists of a nucleus surrounded by fourteen electrons, each with a negative electric charge. The nucleus has a positive electric charge of fourteen times that of the electron, which is why fourteen electrons are attracted to it. That is a familiar kind of configuration. The solar system has eight planets (and some rocks and rejects) in orbit around the central Sun. It is tempting to picture the silicon atom as a tiny solar system with fourteen little electron-planets orbiting the nucleus. But as we will discover, electrons are not little planets; they are in reality something quite new and different.

As our boat sails east and we shrink to ever-smaller size, the world around us changes. Most of the predictable laws governing the lands we travel turn out to be obeyed only on average, and electrons and the other objects we encounter are radically different from the features we are familiar with in the west.

For reasons connected with this, in ways that should become clearer during this voyage, the ability to see smaller and smaller pieces of matter requires particle beams—microscopes, effectively—of higher and higher energy. This means that the frontier of the very small also becomes the frontier of high energy. The important point of high energy in particle physics is the concentration of the energy into a small space, or equivalently into a small number of particles. Because of this, a map of high energies and short distances also informs us about the physics of the very early universe: the hot, dense moments just after the Big Bang. In those first few moments, the energy in any given volume of space was so high that the smallest constituents of matter were laid bare.

To understand all of that, we first need to find out what inhabits this strange new world. What might we find inside an atom? For now, all we know is that the things we find will be very small, and that we need a lot of energy to access them. But where are we going? What strange seas are we sailing, and what laws, if any, apply? The place to start our first expedition is in the first relatively safe harbor we spot in the distance—Port Electron, on the strange shores of an unknown island.

2

The Ocean Wave . . .

From our harbor of Port Electron, we want to chart a course toward the coast we see faintly on the horizon. The local pilot we have engaged is impatient to steer us out of the harbor, through the calm of the bay, and around the patches of choppy water we can see by the harbor's entrance. But our navigator and captain are cautious. Conscious of the challenges ahead, they want to understand what causes the choppiness and how to steer the boat safely themselves. The pilot shrugs and starts talking about particles.

Particle-like behavior is something we are quite familiar with. If you shoot a gun, the bullet will travel onward in a straight line until some force acts to change its direction or slow it down. Sand will trickle through your fingers and form neat piles. These particle-like things will not travel in anything other than a straight line unless they ricochet off something or some other force acts on them to bend their path. They will also stay the same shape as they travel. To properly describe a particle and predict its behavior, we need to know its size, speed, and mass. We think of the molecules in a gas as particles, bouncing off each other, and with that picture we can understand temperature, pressure, and quite an array of interesting and useful behaviors, including convection currents that transfer heat energy around the cabin. Particles also provide a way of transferring information. The letters the crew sent home before we set off on our journey are particles, too, in a sense—discrete packets of stuff traveling in a well-defined path from sender to recipient.

Waves provide a very different way of transferring information and energy. The ship's radio (emergency use only) sends signals back to base, and the ship's microwave heats up the captain's soup. Most of what we know about the world around us comes from waves—generally light waves and sound waves in everyday life, but also radio, X-rays, and more esoteric forms once we bring scientific instruments into play. The physics of waves is in many ways more interesting and more complex than the physics of particles, and gives rise to a richer mix of effects, including the choppy and calm patches the captain saw as we arrived.

To properly describe a wave, we must know its wavelength, frequency, and amplitude. Traveling waves have peaks and troughs that move as they go. But what is actually traveling? The pilot draws our attention to a seagull sitting on the water of the bay as ripples in the otherwise calm sea pass by and lap at the shoreline. The gull bobs up and down as the waves pass but does not otherwise move. Though waves are traveling across the bay to the shore, the seagull, and indeed the water through which the waves are transmitted, only move up and down; they do not travel along the surface. It is only the "up-and-down" motion, the displacement, that is somehow traveling across the bay.

The height of the "up" or the depth of the "down" compared to the surface of the undisturbed bay is what is called the "amplitude" of the wave. Any wave has an amplitude of some kind—the displacement that it causes from the average. An amplifier in a sound system is so called because it increases the amplitude of a wave—it amplifies it, and the sound gets louder.

As the ripples continue—perhaps a dolphin is having a good time nearby, splashing around—then the gull will keep on bobbing up and down. The number of times it bobs in a given period of time is known as the frequency of the wave—the number of peaks or troughs passing a given point in a certain time interval. Usually it is measured in hertz (Hz), a slightly odd unit that should really be called "per second." If the wave in the bay has a frequency of 2 Hz, the gull will bob up and down twice every second.

The wavelength, on the other hand, is simply the distance between two neighboring peaks in the series of ripples. And since the displacement has to travel a distance of one wavelength each time the gull bobs, the speed with which the wave travels across the pond is quite easily calculated by multiplying the frequency by the wavelength.

So if we know the amplitude, the wavelength, and the frequency of a wave, we also know its speed, and that specifies all of its most important properties. How is the behavior of waves any more interesting than the way particles behave?

Well, consider this. Two dolphins are splashing around in different places in the bay, making waves with the same amplitude, frequency, and wavelength as each other, but traveling in different directions. Things might be looking a bit turbulent for the gull. But perhaps not.

If the peaks of two waves arrive at the seagull at the same time, then indeed the bird is in for a bumpy ride. The amplitudes of the waves will add up, and the gull will bob twice as high and dip twice as low. But, depending on how far away each dolphin is from the gull, it might be the case that the peak of one wave arrives just as the trough of the wave from the other dolphin turns up. In this case, the trough will cancel out the peak; or, thinking of it in terms of the water under the seagull, the force from one wave is telling it to move up, while an equal and opposite force from the other wave is telling it to move down. It won't move. The gull can relax. The waves will carry on past it, but it will be left in peace.

Such quiet spots are seen when all kinds of waves meet each other. Radio waves and microwaves, such as those that carry Wi-Fi signals, exhibit them, too.[1] These effects, when waves come together, are known collectively as "interference." When two waves arrive with the peaks of one hitting the troughs of another, they are said to be "out of phase." And obviously, when the peaks come together, they are "in phase." Phase is another important property of waves, but it can only really be defined when you have two waves. Phase differences—such as whether two waves are in or out of phase—have a real physical effect. In our example, the gull bobs up and down, or does not, depending upon the relative phase of the two waves. But phase

1. I think there is such a quiet spot in my office at CERN, the European Organization for Nuclear Research, which is why, in the birthplace of the World Wide Web, I still struggle to get on the Internet sometimes. There are just too many waves arriving at the wrong times and happening to cancel each other out in the vicinity of my desk.

has to be defined relative to something. If there is only one wave, we might decide to define the phase relative to some arbitrary time—say, the moment we first saw the dolphin—but regardless, if there is only a single dolphin making a single set of waves, the gull will bob up and down, whatever the phase of the wave might be. It is only when we have multiple waves with phase differences between them that we see really different behavior. This rather simple fact has surprisingly far-reaching consequences.

This interference behavior is very different from the more familiar behavior of particles. While bullets fired at a seagull from different directions may collide, there is no way that firing more shots could reduce the number of bullets.[2] But making more waves might indeed make its part of the bay calmer.

Waves do other interesting, non-particle-like things. The bay contains a harbor, connected by a narrow channel. All the dolphin-and-seagull action is happening in the bay, and some of the waves impinge on the narrow channel leading to the harbor. What happens?

If waves behaved like particles, then any that were directed accurately enough at the channel would pass through and travel in a straight line across the harbor, leaving most of the surface of the harbor undisturbed. But this is not what happens. The waves hit the channel,

2. I apologize to the seagull. That got unexpectedly violent.

and the channel acts as the source of waves in the harbor—as though a dolphin had actually gotten in there. (This works most effectively if the width of the channel is comparable to the wavelength of the waves, as in that case it looks like a single source of waves, rather than a row of sources.) Waves will spread out from the channel concentrically, across the dolphin-free harbor. This spreading out is called diffraction; it allows waves to go around corners without any bending force being applied. It's another key property that features in the quantum-particle-wave world of the Standard Model.

One important practical consequence of this kind of wave behavior is that there is a limit to the smallest structures they can be used to study. Roughly speaking, effects such as diffraction and interference mean that a wave cannot give us good information about objects that are smaller than the wavelength of the wave. Smaller than that, and things become too blurred and confused. In the case of the harbor channel above, wavelengths much shorter than the width of the channel lead to a tight beam pointing back to the position of the channel. Wavelengths the same size as the channel spread out and fill the harbor; longer wavelengths won't even pass through the gap.

Any setup that can support a wave has an equation behind it—a wave equation, of course—that describes how the wave will work. The surface of the bay we are sailing on is one such system. Another example is the air. A small region of dense, high-pressure air will spread out, compressing neighboring regions, which in turn compress their neighbors, and so on. A high-pressure pulse propagating through the air like this is a sound wave, created when air is compressed somehow, say by a vibrating drum, or your larynx. Electric and magnetic fields form another system, which is how light, radio, and other electromagnetic waves travel. The important point here is that the general behaviors of these systems are similar in some very important ways—including the fact that diffraction and interference occur—because the underlying wave equations are very similar.

Because they will be such a vital navigational aid in our coming voyages, it might be worth taking a moment to examine why equations are so important in physics. We won't need to go into the detailed mathematics, and I won't be writing out any equations explicitly, but there will be several moments when an equation of some kind is so vital to navigating the physical world that we will need to discuss it. An equation in mathematics relates different concepts to each other in an abstract, but completely definite way. When used in physics, the concepts on each side of the equation are physical objects, and an equation relating them gives new insight into how those objects behave, and especially how changing one of them affects the others.

In the current case, a wave equation describes changes in some physical quantity—the height of the water, the pressure of the air, the strength of the electric field. It relates how they change as time passes to how they change with position. Specifically, the wave equation for our bay tells us that if the height of the surface is different at different points in the bay, this implies that the surface will also change with time. Imagine a wriggle from the tail of one of our dolphins that raises a region of water to be higher than its surroundings. This is an unstable situation. The small hill of water created by the dolphin will be pulled down by gravity, and this will spread ripples across the surface as a traveling wave. The wave equation is simply the mathematical description of how this happens. It tells us how differences in the height of the water at different places lead to changes in the height at different times. It can be used to predict how waves will travel and interact—water waves, sound waves, radio waves, or quantum waves.

Our boat heads out of the harbor in a straight, particle-like line, the crew cheered by our pilot's instruction and buoyed by the waves. We now know, and hopefully understand, two distinct behaviors—particle-like and wave-like. They differ from each other profoundly, and it is very hard to see how the two could ever be mixed together. But we are sailing uncharted and dangerous seas, and we should expect surprises. And, to the frustration of some of the more impatient crew members, our pilot is not done yet.

3

. . . Or Particle?

Before we travel on, we need to really understand the medium we are moving in. If we don't do this, the pilot assures us, we will understand little of what we see, and in particular the interior of Atom Land, the target of our next voyage, will be an impenetrable jungle. The coastline already appears closer, though we have barely left port.

What the pilot has to tell us is so weird that he knows we may not believe him, so he urges the captain to drop anchor and assemble the crew belowdecks for a demonstration. After a few moments' preparation, in the pitch-black hold of our ship, the pilot fires a beam of laser light at a screen with two small slits in it. On the other side of the screen is a detector to monitor the light that makes it through the slits.

The first thing to note is that light behaves like a wave. If the slits are narrow enough, the slits themselves start acting as sources of waves. That is, the light diffracts as it passes through the slits, just like the water waves in the narrow channel in the harbor. This indicates that the light has a wavelength that is similar in size to the width of each slit, just as the water waves that diffracted the most had wavelengths similar to the width of the harbor entrance.

Furthermore, we will see a pattern of bright and dark bands with our detector. At each position on our detector, light is being received from two sources—the two slits, like the two splashing dolphins near the harbor. For points exactly halfway between the slits, the light travels the same distance from each slit, and the peaks of waves from each slit will arrive together, in phase. The peaks reinforce each other, as do the troughs, leading to a strong wave and thus a bright light. For any other point, the light travels a different distance from one slit than the other, and the "adding up" is not guaranteed. If the difference in distance traveled is a whole number of wavelengths, the peak from one source will arrive with a different peak from the other, and things will still add up. But if the difference is a whole number plus a half, the peak of one will arrive with the trough of another. Then, the waves are in antiphase. (These are the dark bands, where the peaks and troughs cancel each other out.) The detector will remain dark, just as the seagull rested at peace in the bay.

This seems pretty conclusive. Diffraction and interference are going on, and they only happen with waves. We would not see this behavior with particles. We can even work out the wavelength of the waves, which is not something that makes any sense for a classical particle. Light is a wave. End of story.

But it is not the end. There is a twist. An important twist.

The pilot urges us to look a bit more carefully at the detector that is measuring the light once it has passed through the slits, creating the bright and dark interference bands. The detector in our experiment relies on the "photoelectric effect"; that is, when light hits the detector, the light releases electrons, which can then carry an electric current. The explanation for this behavior lies on the coast of Atom Land, but for now we see that by applying a voltage to the detector, we can make the current flow and detect the freed electrons. This is how we know when light hits the detector, and thus where the bright bands are and where it is dark.

Waves move energy around. That's what makes the seagull move, and that's what releases the electrons in our detector. And with waves, there are two ways the amount of energy can be increased. You can increase the amplitude of the wave, which makes the gull bounce higher. Or you can increase the frequency of the wave, which will make it bounce up and down more often. The same is true with light. The power of a laser can be increased either by making it brighter or more intense, or by increasing the frequency. Frequency for light corresponds to color, so increasing frequency might mean moving from red light to blue light, say.

In our experiment, however, these two different ways of increasing the power have very different impacts on our light detector.[3] One would expect, then, that when the amount of light shining on a photoelectric material, such as our detector, is increased, the electric current would also increase; and this is true under some circumstances. But it doesn't always work like that.

Say, for example, that the light we are using is blue. This means it has a wavelength of 475 billionths of a meter, corresponding to a frequency of 650 terahertz (650 trillion oscillations per second). The light detector registers the light, producing our lovely interference pattern of bright and dark bands, and clearly demonstrating the wave nature of light. If we increase the

3. This difference spurred the development of quantum mechanics, and inspired a breakthrough from Albert Einstein that resuscitated the idea of light as a particle.

power of our blue laser, the intensity of the light registering in the detector also increases. All well and good.

However, we now tune the frequency of our laser. We reduce it, making the light first green, then red. For this particular detector, as the frequency reduces into the red, the electric current suddenly dies away and we can no longer detect the light.

As we reduce the frequency, we are reducing the power of the laser. If this were the waves in the bay we were dealing with, we would be making the gull bounce less often. So it is not surprising that we get less current, though it is a bit surprising that it drops so suddenly.

No matter, we can compensate by turning up the intensity of the light—corresponding to making the gull bounce higher, even if it is bouncing less often.

What we would see is disappointing. In fact, we would see nothing.

Once the frequency has dropped below a certain value (which depends on the detector we have and exactly what it is made of), we have no electric current, no matter how high we turn up the intensity. This is impossible to explain using continuous waves of light. The energy is there—why doesn't it free up any electrons?

This can only be explained if light comes not as a continuous wave, but in little packets, or quanta, of energy, more like the letters the crew sent home than the radio waves we use in emergencies. For light, the packets are called photons. A single photon is a quantum of light. This is the explanation Einstein put in a breakthrough paper of 1905.[4] The energy of an individual photon depends on its associated frequency—blue photons have more energy than red ones. The total amount of energy in the laser beam is the number of photons multiplied by the energy of each photon. When we turn up the power of the red laser, we are increasing the rate at which photons are emitted, but the energy of each photon remains the same, because the frequency of the light does not change.

Conversely, as we turn down the power of the blue laser, we reduce the number of photons, but not the energy of each photon, and so, as Einstein said in his paper, there is indeed "no lower limit . . . for the intensity of the exciting light below which the light would be unable to act as an exciter." Acting as an "exciter" in our case means releasing an electron and thus registering in our detector. This is presumably more elegantly put in the original German, but the result agrees with experiment and is exciting, in any sense. What is meant is that even a laser turned down until it emits only one photon a year would still, eventually, build up the interference pattern of light and dark

4. *Annalen der Physik* 17 (1905) pp. 132—48. See http://einsteinpapers.press.princeton.edu.

bands—one spot at a time. Light comes in discrete packets, as though it consists of particles, but exhibits interference, as though it is a wave.

Put together, the facts that, on the one hand light exhibits wave-like properties such as interference, but on the other hand also comes in discrete packets with an energy that depends on the frequency tell us that it is neither a wave nor a particle, as we know them classically. It is something else entirely. At low intensities and high frequencies, we have entered a new regime of physics and we need a new set of concepts to describe it. Photons are excitations in a "quantum field." The quantum field is the sea that we are sailing on.

The pilot is looking quite smug at this stage and has made a captive audience of the crew. His demonstration has gotten our attention, and gotten us closer to understanding what a quantum field is and how it works. But we need to know more. And he's only too happy to tell us.

4

Traveling in the Quantum Field

A "field" in physics is any quantity that has a value for any given point in space. For example, a magnetic field has a strength at all points near a magnet, which you can measure by the effect it has on small pieces of iron. Earth has a gravitational field defined at any given point, which can be observed by its effect on any piece of matter placed at that point. It keeps our boat firmly on the surface of the sea and makes the rain fall downward from the clouds above. Indeed, without gravity we can't even define "down" and "up." A quantum field takes this idea of a field into the realm of the very small.

Back to the experiment with the laser, the two slits, and the detector—the quantum field can describe what is going on. The magnitude of the quantum version of the electric and magnetic fields tells us how likely we are to find a photon present. This quantum field spreads and travels like a wave, it has a frequency and wavelength, and it can exhibit interference and other wave-like effects, but it is telling us the probability of a particle (a photon) being present at any given place. The energies and the momenta of those photons are determined by the frequency and wavelength of the quantum field. This way, the detector can register individual photons one at a time, but their distribution over time will build up the pattern of light and dark bands that we observe.

The quantum field theory that does this for us is called "quantum electrodynamics"—QED—and was developed by Richard Feynman, Julian Schwinger, and Sin-Itiro Tomonaga in the 1940s. The name is descriptive—the theory treats light as carried by the photon (a quantum) but describes the motion of electric and magnetic fields (electrodynamics). It constitutes the first solid component of what became the Standard Model of particle physics, and we will see a lot more of it on our travels.

As well as describing the apparent contradictions of our experiment, the concept of a quantum field has even more to offer. Electrons are also excitations in a quantum field. As such, they also have wave-like properties. These wave-like properties are seen in interference experiments just like the one we carried out above for photons. It will turn out that these properties are also what is needed in order to understand the interior of Atom Land, when we get there, and the chemistry of the elements.

Quantum field theory also explains the double meaning in the longitude scale of the map we have started to draw on our travels. As we go from left to right, west to east, we also go up in energy, and down in size. This seems odd—high energy means high mass, which usually means "bigger." It is true that in everyday life, heavy things are often (though not always) bigger than light things.[5] But for fundamental particles in quantum field theory, it is the other way around. High energy corresponds to a high frequency, which corresponds to a short wavelength. And as we saw in the harbor, the wavelength determines the smallest thing that can be observed. Thus, to observe smaller objects, more energy is needed. And this means there is a sense in which the objects we discover as we travel east, which have higher and higher masses, are nevertheless smaller and smaller than the objects to the west.

5. This is why on some physics wall charts, heavy particles are shown as big blobs compared to smaller blobs for lighter particles such as the electron.

The incorporation of particle-like and wave-like properties into a new kind of object with the properties required to describe nature is the achievement of quantum field theory.

The pilot has finished his exposition and returns to the wheel as we weigh anchor and get underway. The crew is still absorbing what they have learned. Quantum field theory is very much counter to our intuition about how physical objects should behave, and there is another useful way to try to understand what is going on. Richard Feynman, one of the originators of QED, was a great explainer, and he used an idea called a "path integral" not only to build the mathematics of his theory, but also to describe it to non-specialists.[6]

He talked about particles traveling along all possible routes between two points but carrying with them a rotating "phase," which he visualized as a little arrow. The arrows rotate as the particle travels, and the number of rotations per second is the "frequency" associated with the particle. Rather like our boat sailing one of many possible paths between Port Electron and the approaching shores of Atom Land, with the ship's clock ticking the seconds and minutes as we travel.

Unlike our boat, the particles that Feynman describes are quantum particles, and individually they potentially

6. *QED: The Strange Theory of Light and Matter*, 1985.

travel all over the place, randomly, in all directions. To calculate the probability of a particle actually getting from any place A to another position B, quantum field theory says that all the possible routes between A and B have to be taken into account. Every possible way a particle could leave point A and arrive at point B has to be summed up to get the actual probability of a particle getting there. If that seems weird, well, it is, but it is the way things are, so bear with it. This is where the quantum indeterminacy of very small things enters in. This "summing over all paths" is called a path integral.

The key is that the sum takes into account the direction of the arrows. Remember, the arrows rotate as the particles move, like the hands of the clock on our boat, so for routes of differing length, the arrow will in general be pointing in a different direction by the time the particle gets to B, because it will have had more, or less, time to rotate. The direction of the arrow is like the height of a wave in the harbor. If two arrows point in the same direction, they add up to a single longer arrow. But if they point in opposite directions, they cancel each other out and the sum is zero. This is where the wave-like aspect comes in, because this canceling is just like the peak and the trough of waves arriving at the same time, canceling each other out (and leaving the seagull in peace).

In general, there are so many possible routes from A to B (including ones where particles change mass, and even routes that go backward in time!) that for any given route, there is usually another route that ends up with the arrow pointing in the opposite direction, canceling it out. It is possible to pair up routes like this and show that they make little or no contribution to the final probability of the particle arriving at B. The only place where this doesn't work is for paths that are close to the shortest possible path between A and B. This is the route where the particle undergoes the fewest number of turns of its arrow as it travels, and all the possible routes similar to this one will have arrows pointing in the same direction.[7] Because the arrows are pointing in nearly the same direction, they add up, and the net result of summing over all the paths is dominated by these few paths adding up together strongly, while all the others cancel each other out. This tells us the most likely way a particle will behave, and what the chances are of making it from A to B. If we obstruct the shortest path, for example with the screen with slits in it in our experiment, we have to redo the sum—the path integral—and we get new

7. This is the same effect as minimum height of the land in a valley. On the sides of a valley, adjacent positions have different heights due to the slope, but at the base of the valley—the minimum height—the ground is almost flat and adjacent points have almost the same height. In the same way, paths near the minimum number of turns have a very similar number of turns to each other, so they add up.

behavior, which includes interference patterns, and diffraction and other wave-like effects, exactly as observed. Making the calculations this way gives results that agree with measurements that include not only these wave-like aspects, but also particle-like aspects such as the photoelectric effect.

This is a lot to absorb, and the crew go about their various tasks with thoughtful looks on their faces once the pilot has seen us out of the bay and departed. In the lands we are exploring, the objects we will encounter are not really waves or particles in the way we think of those things in everyday life. But then, why should they be? We are sailing into new territory. We will continue to use the words "particle" and "particle physics," but remember as well that we will not be encountering particles as we know them. They are *excitations of energy in quantum fields*. The quantum field is all-pervasive in our current understanding of the map of physics. It is now the ocean surrounding and connecting the different landscapes of physics we will explore.

The boat is just a boat, of course, and behaves like a big particle, not a quantum excitation. If it goes by some strange route, it isn't going to get canceled out by some other quantum version of itself on which the ship's clock shows a different time. Nevertheless, the navigator is concentrating very hard on taking the shortest path to Atom Land.

EXPEDITION II

Atom Land

Landfall among the atoms – Dead mice, the Sun, and the Standard Model of chemistry – (Don't mention the moles) – A brief trip backward – Attention to the electron, and avoiding the pilot – Into Atom Land – Puddings or planets? – The music of Atom Land – A guitar lesson, shells, and Schrödinger

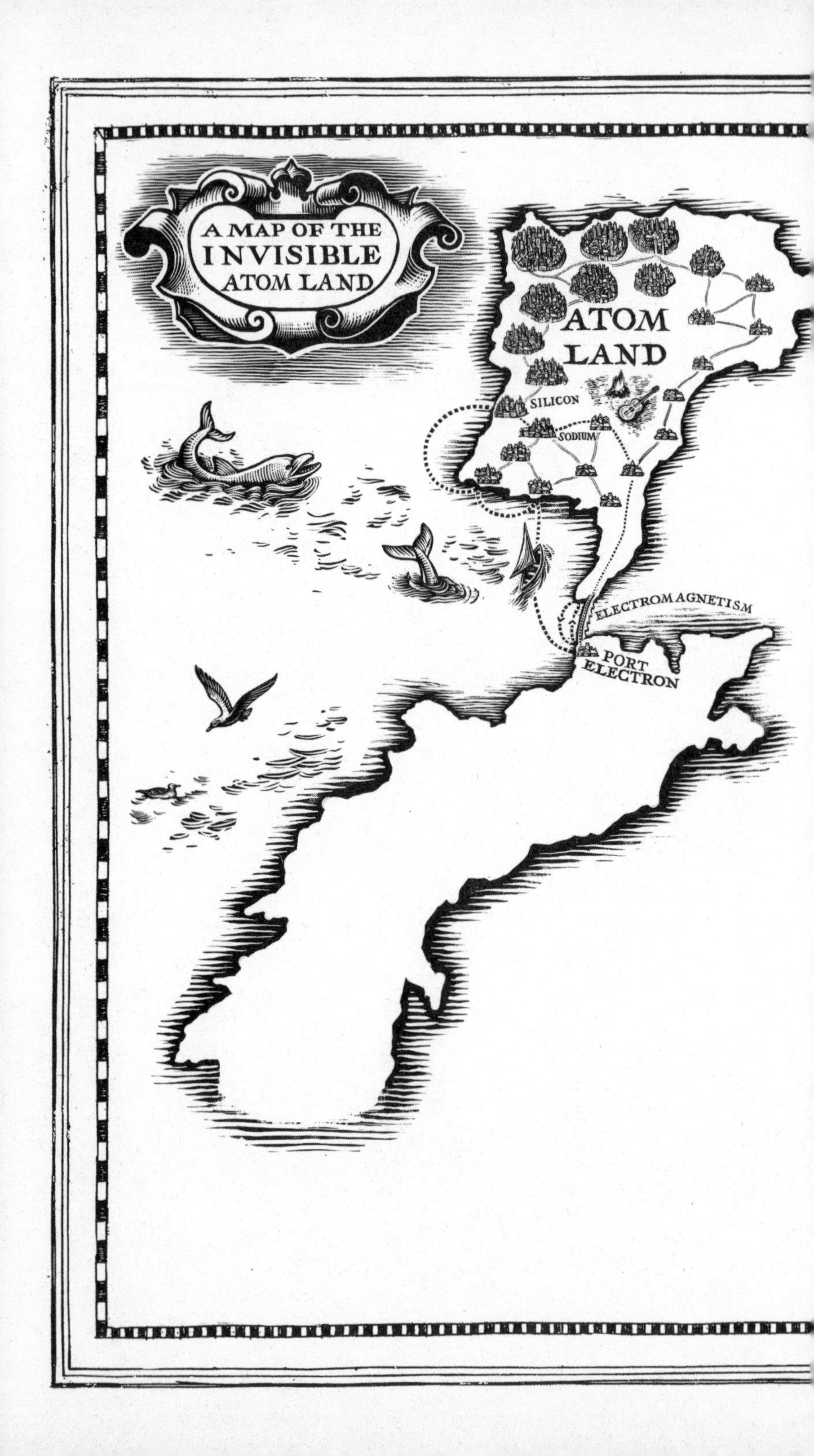
A MAP OF THE
INVISIBLE
ATOM LAND
ATOM
LAND
SILICON
SODIUM
ELECTROMAGNETISM
PORT
ELECTRON

N
W
E
LOW ENERGY
HIGH ENERGY
S

5

Atoms

Having prepared ourselves for the fact that the objects we encounter will be a strange amalgam of wave and particle, we approach the shore of Atom Land with confidence and excitement, eager and, we think, well prepared to explore the interior. We disembark and head off on foot.

An atom is the smallest indivisible fragment of a chemical element. Think back to the fiberglass of our boat, and the silicon the fibers are made of. We already peered briefly inside the silicon atom, observing the nucleus and especially the electrons around it. If we were to break up an atom of silicon, what we would get might be interesting,[8] but it would *no longer be silicon.* Everyday materials consist of different chemical elements, each one of which is a different kind of atom, sometimes bound together as molecules.

The idea that there are indivisible building blocks to matter may date back to the ancient Greeks, but the knowledge that atoms are real was the result of careful exploration over the last two centuries. Much of that exploration was not done directly with high-resolution instruments probing tiny structures, it was done by observing the properties of different materials, identifying how they combine and react with one another, and weighing them precisely. Most of the elements were discovered between 1745 and 1869 by many different explorers using a wide variety of inventive techniques, including tasting, smelling, weighing, or simply observing the properties of various materials and the products of various reactions between them.

For example, several scientists working independently in the 1760s worked out that air contains two major components, one of which would allow flames to burn and make mice more active and healthy, while the other would put out flames and suffocate mice. In the 1770s the mouse-friendly gas—which was also produced by heating mercury oxide—was identified as

8. It is.

the element oxygen. A Scottish student, Daniel Rutherford, guessed that the mouse-killing gas was another element, nitrogen, in his 1772 doctoral thesis.

Astronomy was also employed, being responsible for the observation that there was a new element present in the Sun, not previously known on Earth but identified by the distinctive frequencies of light it emitted. It was named helium, after Helios, the Greek god of the Sun, and was later identified in gases emitted by Mount Vesuvius. In 1895, Swedish chemists Per Teodor Cleve and Nils Abraham Langer noted that the same gas was produced by dissolving certain minerals in acid, and they managed to isolate enough of it that they could measure its atomic mass.

John Dalton, a chemist, physicist, and meteorologist working in the nineteenth century in Manchester, England, conducted a series of enormously careful experiments[9]—combining, reacting, and weighing various gases and other substances—and established that some materials involved in various chemical reactions always combined in fixed ratios. He hypothesized that this was due to the fact that the reaction was actually taking place between *tiny fractions* of each material. These tiny building blocks possessed, he believed, certain well-defined ways of combining and recombining to

9. It is tempting to think that these must have been a welcome break from studying the English weather.

make new, stable building blocks of a new material. Carbon dioxide, for example, could be made from combining two parts oxygen with one part carbon. Water could be made by combining two parts hydrogen with one part oxygen. If you get the ratio right, all the initial materials will turn into the final product. If you get it wrong, you will find that you have something left over.

In 1869, the Russian chemist Dmitri Mendeleev arranged the known elements according to their chemical properties into the periodic table. This is more than just a neat way of arranging things. Grouping the elements by their reactivity and masses in the way that Mendeleev did reveals a pattern that reflects the internal structure of the atoms, and that had predictive power. Gaps in his original table suggested "missing" elements, all of which have since been discovered. It is the "Standard Model" of chemistry, in a sense, and it points rather directly toward the beginnings of the Standard Model of particle physics.

Our exploration of Atom Land has shown us a complex and beautiful array of features, the building blocks of everyday materials. They exhibit some fascinating behavior, which we would like to understand better. But everywhere we go in Atom Land, everyone we ask tells us that to understand the economy and ecology of the place, to grasp how its denizens really interact with each other, we must return to Port Electron.

6

Going Subatomic: The Electron

A short sea trip later, retracing our previous path, we pull into the bay again at Port Electron. Despite the lessons from the garrulous pilot, our brief visit to Atom Land has shown us that we need more answers from here before heading into the interior. We disembark, hoping to avoid meeting the same pilot, and spread out to explore. This is what we find out.

The electron was the first subatomic particle to be discovered. These tiny objects were first observed as beams of so-called "cathode rays," a strange radiation emitted by metals when they are heated. Some thought the rays were made up of particles, while others thought they were waves in the ether. Two decades after their initial discovery, J. J. Thomson, working in Cambridge in the UK in 1897, appeared to settle the matter in favor of particles.

Particles have a definite mass and a definite electric charge, which also means that the ratio of the mass and the charge has a definite value. To prove that cathode rays were made of particles, one thing Thomson needed to do was to show that this ratio was always the same, regardless of what material was used as the source of the cathode ray. This would meet the criteria for calling them "particles."

The first key evidence is the fact that cathode rays are deflected in electric and magnetic fields in just the way that would be expected for a beam of charged particles. No wave known at the time carried charge, so this might be considered strong circumstantial evidence in favor of the particle hypothesis.

The second piece of evidence Thomson acquired came from finely balancing the electric and magnetic fields he applied to a beam of cathode rays as it traveled through a vacuum. He was able to arrange it so that the forces from each field canceled each other out completely, so there was no net force. From this setup, the speed of the beam can be worked out.[10]

10. The magnetic force on a particle depends on both its charge and its speed, whereas the electric force depends only on its charge. So when these are made to be equal, with a bit of algebra the unknown charge can be canceled out and the speed can be calculated.

Finally, once the speed is known, the magnetic field can be turned off, and the amount the beam deflects in the electric field allows the ratio of the charge and the mass to be determined.[11] As Thomson observed, the ratio for the electron is about two thousand times higher than it is for the hydrogen ion, which is a single proton—the lightest particle known at the time. This meant that either an electron carried enormously more charge than a proton, or it had much less mass.

There are lots of ways of working out which of these options is true. Probably the most annoying way, which the pilot would doubtless be eager to demonstrate if we hadn't managed to avoid him this time, is to suspend small charged spheres in an electric field. The electrostatic force on the sphere depends both on the strength of the electric field and on the electric charge that happens to be on the sphere. If the sphere is stationary—not falling or rising—this force must be exactly canceling out gravity, which depends on the mass of the sphere. So if the field strength and the mass of the sphere are known, the charge can be calculated. Doing

11. Knowing the speed is crucial, because that tells you how long the electric force acts on the electrons as the beam passes between the plates. The impulse from the force deflects the beam by an amount depending on the charge (high charge means a larger force, so a larger deflection) and the inverse of the mass (high mass means more inertia, so smaller deflection). The result is that the deflection depends on the ratio of the charge and the mass.

this many times shows that the charge is always a multiple of a small unit, which we call *e*. Spheres can carry a charge of *e*, or two or three times *e*, or a hundred times *e*, but never a half *e*, or any fraction. This *e* is the charge of the electron.[12]

The result of all this evidence is that there is a tiny particle, the electron, with a definite mass and definite charge. Since electrons are so much smaller than atoms, it is a sensible guess to assume they are present somewhere inside atoms, before being split out to form cathode rays.

At this point we are truly ready to return once more to Atom Land and begin the exploration of the interior.

12. This is a very painstaking experiment, even more so when carried out with oil drops instead of standard spheres, as it was originally in 1910 by the American physicist Robert Millikan. The experiment is still carried out, largely without success, in far too many undergraduate physics laboratories by squinting and impatient students.

7

Nuclear Options

The knowledge gained in Port Electron allows us to land again on the shores of Atom Land with confidence, ready to understand some of the features we find there. But there must certainly be other stuff inside the atom, in addition to electrons. Electrons are so light compared to atoms that something has to account for most of the mass. Also, atoms are electrically neutral, so there must be something in there carrying a positive charge to balance the negative charge of the electrons. What might this other stuff be, and how are electrons distributed among it?

The first tools needed to explore further into the territory of the atom were provided by the discovery of radioactivity. Atom Land lies somewhat to the east; to go there and to explore inland, we need higher energies than our gentle cathode rays can provide. Luckily, for reasons that will become clearer later, some elements naturally give off radiation that has much higher energy than anything seen so far. This is the vehicle that will allow us to penetrate the terrain of Atom Land.

One of the most common forms of naturally produced radiation consists of "alpha particles."[13] A landing party of Hans Geiger, Walther Müller, and Ernest Rutherford (in Manchester again) made the most decisive early discovery of the interior of Atom Land. They used a beam of alpha particles, produced in radioactive decays of the quite newly discovered element radon, to bombard gold atoms. The relatively high energy of the alpha particles meant that they should in principle have the resolving power to see tiny, subatomic features. The idea was that the alpha particles would be deflected by such features inside the gold atoms, and by analyzing the angles at which they scattered, and how often they did so, details of the internal structure could be worked out.

No one had been able to look inside the atom before, and there were different hypotheses as to what might be going on in there. One model for the internal

13. The other common forms are beta particles, which are electrons, and gamma rays, which are photons.

structure of atoms, apparently favored by Rutherford himself, was that electrons were distributed throughout the atom like raisins through a plum pudding. Surprisingly, plum puddings don't contain any plums,[14] but this has nothing on the surprise that Geiger, Müller, and Rutherford got when they fired a beam of alpha particles at gold foil. In a plum-pudding atom, the alpha particles should have smashed through with only minor deflections. But while most of the alpha particles passed through almost unperturbed, some of them actually bounced right back again, and others were deflected through much bigger angles than should have been possible for a diffuse, pudding-like distribution of matter inside the atom. Rutherford famously described this result as like firing a 15-inch shell at a piece of tissue paper and having it rebound to hit you.

This remarkable message from the interior can only be explained if the overwhelming majority of the positively charged mass of the atom is concentrated in a volume thousands of times smaller than the atom itself. This, we now know, is the atomic nucleus. The vast majority of the mass of any atom is concentrated in about a thousand-trillionth (10^{-15}) of the volume of the atom itself. This concentration is why it can bounce alpha particles back the way they came. The basic structure of every atom is a very heavy nucleus

14. Apparently there's a good historical raisin for this.

surrounded by a cloud of lighter electrons. The next part of our journey will explore how those electrons are bound to the nucleus, and the far-reaching consequences of that.

8

The Source of Chemistry

Our explorations so far tell us that the vast majority of the mass of an atom is concentrated in the nucleus. The much lighter electrons buzz around the nucleus, constrained to stay in the neighborhood by the electromagnetic attraction between the negative charge they carry and the positively charged nucleus. This is reminiscent of a mini–solar system, with lighter planets in orbit around a more massive star in the center. However, we already know that electrons are not classical particles. This is another point at which their quantum-mechanical nature makes a huge difference. In this case, it dictates the way atoms bind and react to form molecules and compounds, and explains the structure of Mendeleev's periodic table itself.

The reactivity of the different elements depends upon how tightly bound to the nucleus the electrons they contain are. As we tour Atom Land and visit the atoms of various different elements, we find they contain different numbers of electrons—enough to balance the positive charge of the different nuclei. But we also find that these electrons cannot have any old arbitrary energy. They have a specific set of binding energies characteristic to the atoms of each element. These characteristic energies are what determine the ability of the atom to form molecules and other associations with neighboring atoms. They are responsible for the whole of chemistry, and everything that follows from it. Like any curious explorer, we need to understand how this all works: What fixes these energies?

An electron of a particular energy has a particular wavelength associated with it, as we saw on our previous voyage. When traveling around, freely crossing the oceans of our map, electrons can have any wavelength, and therefore any energy, with no restrictions. But when they are confined within Atom Land, bound close to an atomic nucleus, this is no longer true. The fact that only certain energies are allowed implies that only certain, very particular, wavelengths are allowed.

This—the fixed wavelengths—is where we can start to understand what is going on with the electrons. There are other situations where only a few special wavelengths are allowed. One example is the harmonics on a guitar string. Luckily, we have a guitarist on our boat who will help illustrate this. In a clearing in one of the forests of Atom Land we make camp, build a fire, and seat ourselves around it to hear what she has to say, as the dusk draws in and the tiny electrons buzz around the treetops above us.

Each note that a musical instrument makes corresponds with a particular wavelength of sound. A guitar string of a certain length will make a particular note, determined by the fact that an exact number of half wavelengths corresponding to that note have to fit into the space allowed on the string. The ends of a guitar string, at the bridge and the nut (or at a fret when the guitarist holds the string down), are fixed. They can't vibrate like the rest of the string. So a wave on the string must have fixed points at each end, points at which the amplitude of the oscillation is zero.

This has the consequence that not all wavelengths work. A wavelength as long as the string is OK—it would have a fixed point at each end, and another fixed point in the middle, with the peak and the trough a quarter and three-quarters of the way along, swapping sides as the wave bounces back and forth.

A wavelength twice the length of the string also works, with the middle of the string oscillating up and down. This would actually be the bass harmonic of the string, the open note a guitar plays. The important point is that any wavelength that doesn't allow stationary points at each end of the string is forbidden.

That is what happens with electrons, too, when they are confined close to an atomic nucleus. The limits of the electrons' distance from the nucleus are like the bridge and nut of the guitar—they define fixed points that the electron cannot go beyond and where the wave associated with the electron is stationary. This means only certain wavelengths are allowed, and that in turn means that only certain energies are possible, and that, in turn, explains the peculiar structures in which electrons are bound to the nucleus of the atoms we have encountered.[15]

There is a final piece of information we need to make sense of the emerging internal arrangements of Atom Land. There is a definite list of allowed energy levels for electrons bound inside an atom—the harmonics of their orbits around the massive central nucleus. But one might expect that the most stable situation for an atom is that all the electrons sink to the lowest energy level; all of them play the bass note. This is not what

15. The equation that describes these waves is the Schrödinger equation. Less famous than his cat, but much more useful.

we see. Each energy level allows only two electrons to occupy it, and is then full. The NO VACANCIES signs go up for any further electrons, which will then have to make do with the next-lowest energy level,[16] which can then only take two of them, pushing the rest still higher, and so on. An atom in its lowest-energy state has all its electrons in the lowest available levels, with all the higher levels empty. So back to silicon—the fourteen electrons are in the seven lowest energy levels, two of them in each. Sodium has eleven electrons. They will fill the lowest five energy levels, leaving the sixth one half-filled; one electron and one vacancy.

In this way, the atoms build up "shells" of electrons, with energy levels inside the shell filled and those outside empty, with sometimes a vacancy on the edge. This intricate structure of electrons and energy levels determines the size of the atom, and its propensity to react and form molecules.

There are a lot of questions that can be asked here. For example, why two electrons per energy level, not just one—or as many as we want? We do not yet know. But it is difficult to overstate the impact of the discovery itself. The fact that the energy levels are distinct and different for each type of atom, and even for different molecules when the atoms bind together, provides a

16. This unfriendly behavior is the "exclusion principle" discovered by Wolfgang Pauli, one of the first explorers of our landscape, who we will meet again before too long.

way of identifying the constituent parts of materials without touching them.

When electrons jump around between the different energy levels, they emit or absorb characteristic amounts of energy as photons of light. The science of measuring and identifying these is called "spectroscopy," and it is the reason we know what the Sun, the other stars, and the dust between them are made of. Because only certain discontinuous energy levels are allowed, only certain jumps in energy are allowed. So only certain energies of photon can be absorbed or emitted, and these show up either as dark lines in the spectrum of light passing through a material, where those wavelengths of photon have been absorbed, or as bright lines in the light given off by a material when it is heated, such as the characteristic yellow lines in a sodium lamp. The exact yellow, when measured precisely in a spectrometer, is enough to identify sodium as the main component of one of those lamps. Similarly, the lines in the spectrum from other materials allow us to see what kinds of atom are present. This explains the "distinctive frequencies of light" responsible for the discovery of helium in the Sun, for example.

The electronic structure of atoms and molecules—the detailed geography of Atom Land—was discovered at energy scales of a few hundred or thousand electron volts.[17] It was crucial in establishing the quantum nature of electrons and photons, as well as telling us what elements are present in the stars and dust of distant galaxies. It provides much inspiration, and information, for further exploration. The theory of how electrons and photons interact—QED—was the first part of the Standard Model of particle physics to be developed, and precision atomic physics measurements played a critical role in that development, as we will see as we voyage onward.

Atom Land, and the quantum theory we picked up along the way, is the point of departure for the far reaches of particle physics. The next step of that journey takes us back once more toward the island on which we first landed, the island containing Port Electron. For as we reach the south of Atom Land, we discover there is a bridge, as well as a car rental shop. We cross the bridge, rent a car, and head off by road to explore the new territory in the hinterland of Port Electron.

17. One electron volt (eV) is the energy of motion that an electron acquires if it is accelerated through one volt of electric potential. So a 12-volt battery can accelerate an electron to an energy of 12 eV. To get to millions of eV would require several hundred thousand such batteries. Or the nuclear power of one decaying nucleus.

EXPEDITION III

The Isle of Leptons, and Roads Onward

Hitting the road – Maxwell – A unification of forces, with the power to travel – Plus ça change, plus c'est la même chose *– Relativity, quantum mechanics, and a powerful new vessel for the journey south – Solving an old puzzle and discovering a new world – Road trips to the east*

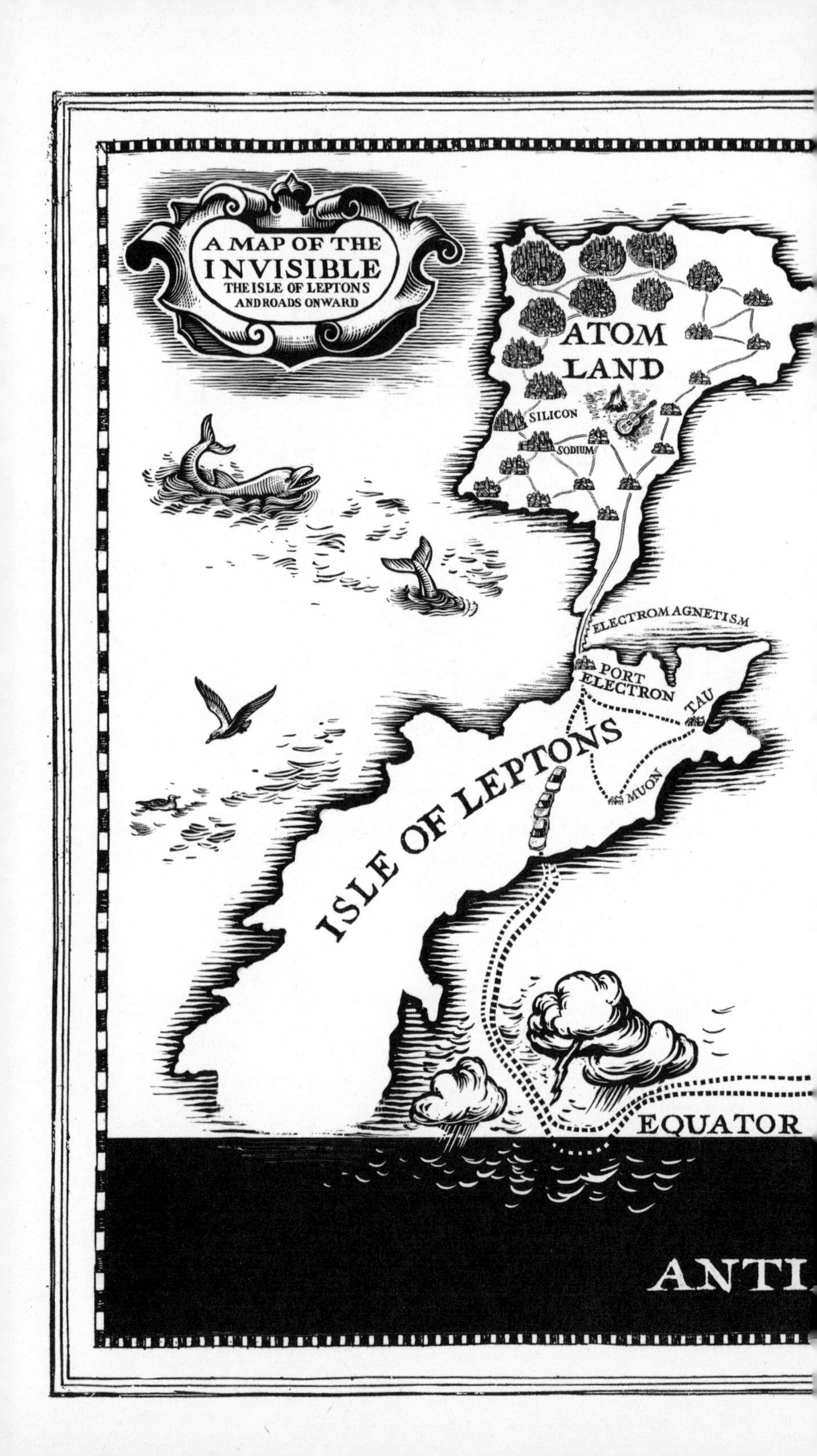

A MAP OF THE
INVISIBLE
THE ISLE OF LEPTONS
AND ROADS ONWARD
ATOM
LAND
SILICON
SODIUM
ELECTROMAGNETISM
PORT
ELECTRON
TAU
MUON
ISLE OF LEPTONS
EQUATOR
ANTI

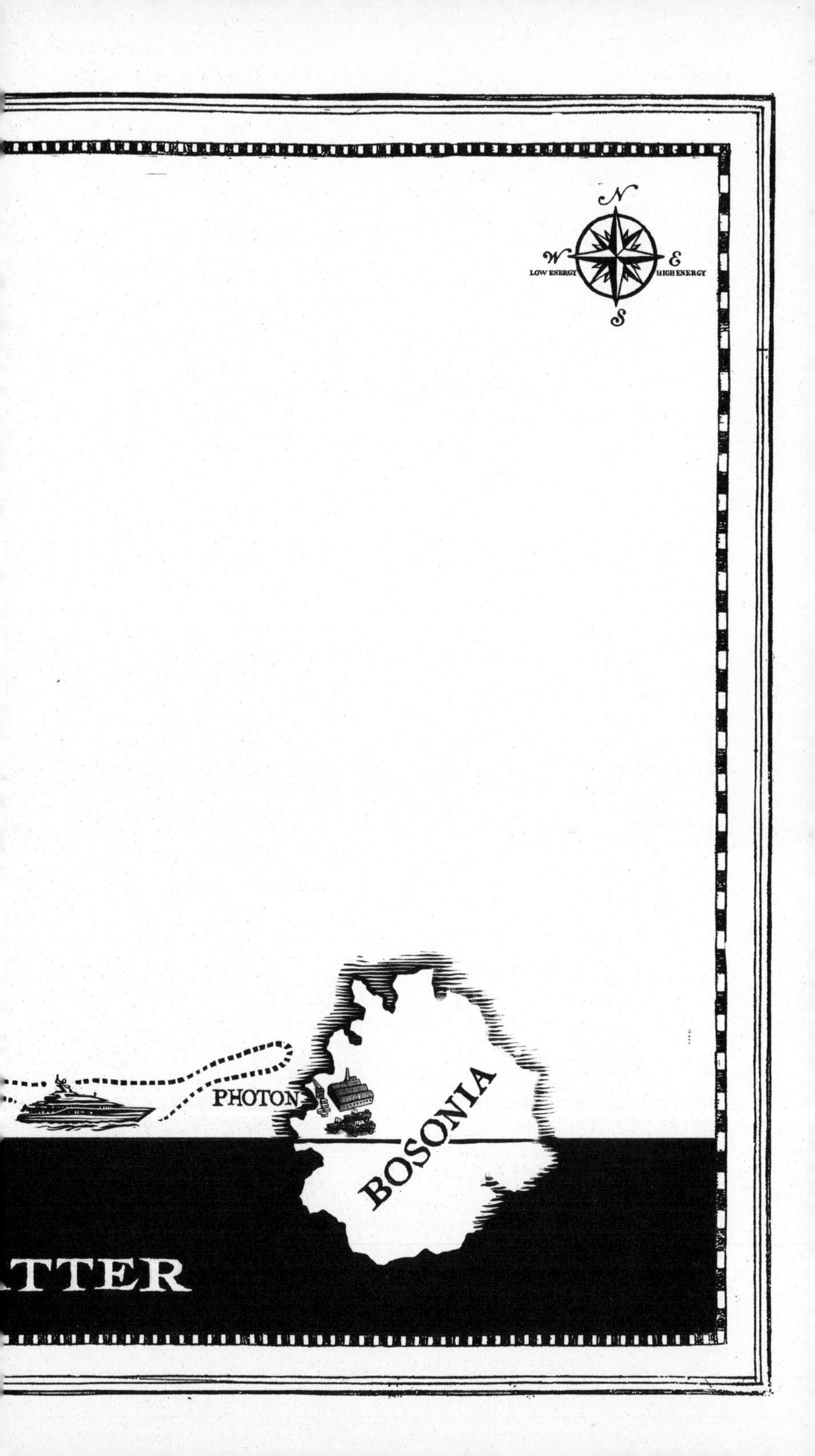
N
W
E
S
LOW ENERGY
HIGH ENERGY
PHOTON
BOSONIA
TTER

9

Electromagnetism

The force connecting the electron, the atomic nucleus, and many of the other features on our map is electromagnetism. Electromagnetism is described by QED, the quantum field theory of Feynman, Schwinger, and Tomonaga that we have encountered already. It plays a vital role in holding Atom Land together, and provides the very bridge we crossed to get here, as well as the road network we are now traveling. The electromagnetic force is carried by the photon, the quantum of light that we have already come across. As we explore the island for which Port Electron was our first point of entry, we will have time to look in more detail at how it works. This will be a surprising exploration, and one that will change the way we see space and time.

On our expedition into Atom Land, we saw that electrons are bound to the nucleus in a complex series of energy levels, because they are quantum particles with negative charge and the nucleus has a positive charge. To get far enough eastward to see the nucleus, we needed alpha-particle energies of a few million electron volts. To see the electrons around the nucleus, less energy is needed—a few thousand electron volts is enough. This is roughly the "binding energy" of the electrons in a typical atom.

Binding energy is an important concept that will feature often on our map. It is the difference in energy between separated particles and the energy of the same particles stuck together. It is always present in any composite object. It means that to pull the object apart, energy has to be added. You can even think of the launch of a spacecraft from a planet in this way. The spacecraft and planet are a bound system. To separate them, you need to put in a lot of energy, released from rocket fuel, to reach escape velocity. Likewise, to remove electrons from an atom—to ionize it, or eventually create a plasma where electrons and ions buzz around freely—energy must be added. If you want to release one of the most tightly bound electrons from an atom, you have to give it energy; otherwise the nucleus will cling to it.

This binding energy determines the longitude of Atom Land on our map. The attractive power of electromagnetism is the force responsible for the binding.

In fact, the interaction we now call electromagnetism is the unification of two forces previously thought to be distinct. There is the electrostatic force, by which two electrically charged objects will attract or repel each other. If they are both negatively charged, or both positively charged, they will repel each other. If one is negative and one is positive, they will attract each other. Most materials are electrically neutral, because the positive charge of the atomic nuclei exactly balances, and so cancels out, the negative charge of the electrons. However, rubbing a balloon on your hair, for example, can transfer some electrons from one to the other, leading to an imbalance of charge and therefore an electrostatic attraction.

In addition to the electrostatic force, if the two electrically charged objects are in motion relative to each other—say electrons carrying a current in opposite directions in two wires—they will also experience a magnetic force, which depends not just on their charge, but on their speed.[18] The magnetic force field of Earth will bend electric currents, and is caused by the flow of currents inside the core.

18. Recall that it was by balancing these two forces that J. J. Thomson discovered the electron.

There are many questions about these forces, though. Both the electric and magnetic forces depend on the electric charge, so surely they must be related to each other. But how? And is the force between the charged particles instantaneous, or does something carry it? If so, what, and how fast does it travel? How does that force behave when the distance between charges changes, or when the charges are moving, or spinning?

These are not obscure details. They go to the heart of physics. And the importance of having a good understanding of them lies not just in the fact that (as usual) we want to know what is going on, but also because the detailed behavior of electrons under this force is the key to much of the natural world, as well as the modern technological world.

Leaving aside the biology for now, the bilge pumps of our little boat chugged away using an electric motor. On our journey we are navigating using a compass. As we do so, I am making notes on a laptop stuffed full of electronic technology. When I've finished my account I will email it to my publisher using Wi-Fi or maybe a cell phone signal. Anyone coming after me will read the words with the help of light, unless they have a braille or an audio version. All of those things, even light itself, are understood in terms of the relationships between electrical charges, and electric and magnetic fields.

All of this is electromagnetism, and the original equations that unified the electric and magnetic forces, that led to QED, and that explained optics, radio, and Wi-Fi are Maxwell's equations, published by Scottish physicist James Clerk Maxwell in the *Philosophical Transactions of the Royal Society of London* in 1865.[19]

Maxwell's equations relate magnetic and electric fields to each other and to electrical charge and current. They encapsulate the fact that electric and magnetic fields make electric charges move; that electric charges cause electric fields, but that there are no magnetic charges; and that changes in magnetic fields cause electric fields, and vice versa. The equations specify precisely how all this happens. They also encode the fact that the only way to change the amount of electric charge in a given volume is to have an electric current take it away or bring it in. Charge never just vanishes, or appears from nowhere; it is conserved.

The equations seem to have been very much a bottom-up affair, in that Maxwell collected a number of already-known laws that were used to describe various experimental results. For example, Michael Faraday, working at the Royal Institution in London

19. The original manuscript was submitted in 1864, but, in a situation familiar to scientists everywhere, it was then held up in peer review. There's a letter, dated March 1865, from William Thomson (later Lord Kelvin) saying he was sorry for being slow, that he'd read most of it and it seemed pretty good ("decidedly suitable for publication").

in 1831, had discovered that moving a magnet through a loop of wire caused an electric current to flow—an effect known as electromagnetic induction. Maxwell built Faraday's law of induction into what was effectively a "grand unified theory" of electricity and magnetism, encapsulating the way charges attract one another and how a magnetic field is generated by an electrical current. While collecting and unifying existing laws such as Faraday's law of induction, Maxwell also introduced one new term of his own. This term describes the way in which a change in an electric field gives rise to a magnetic field, *even if no electrical charges or currents are present*. Maxwell fitted this with the other equations into a unified framework. What is amazing is how much that framework then reveals, in terms of deep physical principles and rich physical phenomena.

Crucially, the equations show that electric and magnetic fields can exist even in the absence of electric charges. A changing electric field causes a changing magnetic field, which will cause more changes in the electric field, and so on. Mathematically this is expressed in the fact that the equations can be rearranged and combined to get a wave equation—a formula to describe a traveling wave, of the kind we met before when watching the seagull in the bay. Because the electric and magnetic fields can support traveling waves, they can carry energy and information.

The speed of these waves can be obtained from the equation, and it is 300 million meters per second—the speed of light! These waves are, in fact, light—electromagnetic radiation; in quantum language, they are photons. They come in many different forms—visible light, radio, Wi-Fi, X-rays, and more. All interact differently with matter—being absorbed and reflected differently by different materials—but these differences are entirely due to the fact that the wavelengths—the distance between successive peaks of the wave—differ. Photons will stay with us as we travel from the lowest energies in the west to the highest in the south; they are our road network on the map that connects everything that carries electric charge. They may appear very different sometimes; visible light does not seem much like a gamma ray. But from the point of view of Maxwell's equations, and of QED, they are all waves in the electromagnetic field.

10

Invariance and Relativity

In our exploration of the landscape of the invisible, equations are a key resource. They relate different objects in the landscape to one another, and give new insight into how those objects behave—as we already saw that wave equations do. Nowhere is this more true than on the roads we are traveling now. Maxwell's equations are such a powerful new resource, they will reward some deep interrogation to see what else they contain.

The equations work in three dimensions, and they relate fields pointing in different directions to one another. So the electric field in the north-south direction depends upon what the magnetic field in the east-west direction is doing, for example. Maxwell wrote it all out component by component, direction by direction, in twenty separate equations. Maybe that's why it took Lord Kelvin a while to read the paper.

There is a more elegant way of writing down the same information, and this also reveals important features that will be an essential instrument for our navigation of the map of physics. The equations can be written down simply, in just four short lines,[20] in terms of mathematical objects called vectors.

A number is a basic mathematical concept that can be used to describe the size, or value, of something. The weight of the vehicle we are traveling in, for example. The temperature of the engine as it struggles up a hill. A vector is a mathematical concept that can describe objects that have both a size and an orientation, like an arrow. Velocity is a vector, for example. Instead of saying how fast our vehicle is traveling in terms of a north-south component and an east-west component, we can give a vector. The length of the vector is our speed, and the angle specifies our direction. Similarly, an electric field has a magnitude and a direction, and can be described by a vector.

As well as being economical with ink, expressing the equations in terms of vectors makes it obvious that

20. This is the way they appear at the foot of the statue of Maxwell near the Royal Society of Edinburgh.

they possess a certain symmetry; like a sphere, they are the same from any angle. If I rotate the directions of the vectors so that north becomes east, or southwest, or whatever, so long as I rotate *all* the directions, all the axes, together, nothing changes. The same equations still work. A physicist, or a mathematician, would say the form of the equations is "invariant" under rotations.[21] If we turn from driving east to driving north, Maxwell's equations stay the same.

Looking for invariances and symmetries like this is one of the surest guides to finding a sensible route around our map of the invisible. And there is another invariance hidden in Maxwell's equations, in addition to rotational invariance. The equations stay the same if I change speed. It is not at all obvious that this should be the case. For example, the equations relate how moving electric charges (a current) create a magnetic field. If I change my speed, I change the apparent speed of the current. I could even catch up to it, so that from my point of view there is no current! So what do Maxwell's equations tell me about the magnetic field in this case?

On our evolving map, the road connections represent electromagnetism. So let's perform a test, or a thought experiment. A series of cars is passing us at

21. This is actually why they save on ink—when we clump the components together into vectors and write the equations that way, the absolute directions of the vectors don't even appear in the equations. The physics does not depend on them.

50 mph, each carrying a large box of electrons, and thus a negative charge. The stream of cars is therefore an electric current, so Maxwell's equations tell us we should see a magnetic field due to that current, as well as an electric field due to the charge of the electrons. We do, and we can measure them both.

Now, imagine we accelerate to a speed of 50 mph, in the same direction as the series of cars. We are now traveling with them. Relative to us, they are stationary, so there is no longer an electric current. And so there should be no magnetic field.

Has physics really changed just because we sped up? From the point of view of someone by the roadside, there is still a current, and so according to Maxwell's equations nothing changed—there *is* still a magnetic field. Do we each need our own version of Maxwell's equations? What about a third person moving at a different speed, maybe traveling in the opposite direction at 50 mph, so they see a stronger current?

The answer is that we all use the same version of Maxwell's equations. They still work, because they are invariant under changes of speed. Indeed, for us the magnetic field will disappear, because the current vanishes, but the electric field changes subtly to compensate, and in the end the relationships between motions and forces on any electric charge all stay the same in a way that appears almost miraculous but is entirely prescribed by Maxwell's equations. The electric and magnetic fields are said to be "covariant"—they vary together in a way that keeps the form of the equations invariant. We only need one version of physics, no matter how fast we may be moving relative to each other.

This has very far-reaching consequences. Remember, Maxwell's equations could be used to get the wave equation for electromagnetic waves, and therefore the speed of light. If the equations are the same no matter how fast we travel, then *so is the speed of light.* The speed of light is an invariant.

The fact that the speed of light is always the same for everyone, no matter what their own speed is, is a founding principle of Einstein's relativity. Relativity is needed to describe the fast-moving and high-energy particles we will meet on our travels, and it also contains the fact that energy (E) and mass (m) are related by the famous equation $E = mc^2$, where c is the speed of light—the speed of the waves originally derived from Maxwell's equations.

This is a lot to get out of one set of nineteenth-century equations. By keeping the electrons bound to the nucleus, they are holding Atom Land together. All particles that have electric charge can pass photons back and forth, either attracting or repelling each other—an important road network connecting many of our islands of knowledge. In addition to all this, the equations have shown us a tool kit of general rules and ideas, like invariance and relativity, that will help us to explore further.

11

The Good Ship *Dirac*

Exploring Maxwell's equations has filled in some of the physics behind QED and built a road network that bridges Atom Land to the electron. It will allow us to travel through the interiors of the islands we discover on our voyages. It has also brought Einstein's relativity into view. But Maxwell's equations are not quantum mechanical, and we already know that without quantum mechanics, Atom Land makes no sense. There is a huge and unexpected reward to be had by looking at how relativity and quantum mechanics come together in QED.

We've seen the wave-like nature of the electrons around atoms, which is an essential feature of chemistry and physics. This wave-like behavior has to be governed by some kind of wave equation, because that's how waves work. The simplest equation that can be used to do this was derived by Erwin Schrödinger in 1925. It takes the classical relationship between the energy of a particle and its momentum[22] and by some ingenious sleight of hand makes a wave equation out of it. The way this is done is to redefine the energy and momentum of a particle as being properties of an underlying quantum "state," a new kind of object that contains all possible information about the particle. The energy is related to how the state changes with respect to time, and the momentum with how it changes with respect to distance.

But this is no longer good enough. We need to take relativity on board.

It is a little as though we have an excellent little boat for chugging around the coastlines of our islands, hugging the shore of Atom Land. But we need something more powerful and heavy-duty to take us farther. We need a bigger boat. That's where the Dirac equation comes in.

The classical, pre-relativity relationship between energy and motion dates back to Isaac Newton and others in the seventeenth century. It states that the

22. Energy is equal to the square of the momentum divided by twice the mass, $E = p^2/(2m)$, which is equivalent to the possibly more familiar expression for kinetic energy $E = ½mv^2$, where v is the velocity.

kinetic energy—the energy of motion—is one half times the mass times the square of the speed. After Schrödinger's trick, this becomes a wave equation that tells us how a quantum particle moves. That equation works very well in predicting the behavior of electrons and other particles, including many subtle quantum effects—for example, the energy levels of the electrons around an atom that we already explored. But the energy-and-momentum relationship we started with is valid only at speeds much slower than the speed of light. It does not incorporate the lessons of relativity. So as we explore the behavior of particles at higher energies (and higher speeds), we need to improve Schrödinger's equation to take into account relativistic effects.

Taking a recipe similar to that used by Schrödinger to obtain his wave equation, the obvious thing is to use Einstein's new relation between energy and momentum and make the same replacements. That is, the energy tells us about the time dependence of the quantum state, the momentum tells us about the dependence on position, and we should get a new wave equation that works even at speeds close to the speed of light.

The complete form of Einstein's $E = mc^2$ includes the momentum of the particle as well as its energy and mass, and relates the square of the energy to the square of the mass and the square of the momentum.[23] And that presents a problem when we try to turn it into a quantum wave equation. Equations involving squares generally have two solutions. If I know that the square of some number is four, what do I know about the number? It might be two, since the square of two—two times two—is four. But it could also be negative two, since negative two times negative two is also four. Likewise, $-E$ multiplied by $-E$ is the same as E multiplied by E. It would seem that our new equations allow negative energy particles. That is not an easy thing to make sense of. What is a particle with a negative energy, or even negative mass? This turns out to be quite an important question.

Remember that to get the right answers about quantum particles—how they move, where they go, how they interact or bind together—we have to include all the possibilities for how they might move, and where they might end up. Only then do we get the proper mix of wave-like and particle-like behavior that is seen in nature. This means we can't pick and choose solutions to our wave equation. All possible solutions have to be taken into account, and that means we have to allow electrons to have negative energy—

23. If you want to know, the full expression is $E^2 = m^2c^4 + p^2c^4$, where E is the energy, m is the mass, and p is the momentum. For zero momentum, this reduces to $E^2 = m^2c^4$, and taking the square root of both sides gets us the familiar $E = mc^2$. And possibly $E = -mc^2$, too (see text).

lots of it. And we don't see those electrons around in nature.

Worse, those negative energy states would allow positive energy electrons to sort of sink into them and vanish. The number of electrons would not be conserved—very bad news for conservation of electric charge, and totally incompatible with Maxwell's equations.

None of this matches the way electrons actually behave. So, at least for electrons, the default recipe for making a quantum wave equation fails. We do indeed need a bigger boat.

The negative solutions are allowed because the energy is squared in the equation. What is really needed is an equation where energy appears only once, not squared.[24]

So, to summarize the problem: To carry us any further in understanding particles, we need an equation that is consistent with relativity, but in which the energy and momentum are not squared—something like, energy is equal to *A* times the mass plus *B* times the energy, where *A* and *B* are some unknown numbers we need to work out. This is the kind of thing we do in mathematics very often. Find *A* and *B* and we're on the way.

24. This is true in Schrödinger's equation. But because of the way relativity mixes up energy and momentum (and space and time), the momentum must appear to the same power as the energy. In the Schrödinger equation, the momentum appears squared, which won't do.

The problem is, there are no numbers that work if we substitute them for *A* and *B*. To make the equation work, you need *A* and *B* to be objects that don't "commute." This doesn't mean that *A* and *B* work from home or live over the shop. "Commuting" in this context means that *A* times *B* is the same thing as *B* times *A*. That's true for all numbers. The multiplication tables learned at school all show commutation, in that if we know five times six is thirty, we know that six times five is also thirty. Even imaginary numbers such as the square root of negative one, which you might think would be weird enough to do just about anything, commute and therefore don't help us toward the relativistic wave equation we need.[25]

At this point it might be tempting to give up and leave the rest of the map blank. Maybe the behavior of subatomic particles is just not susceptible to mathematics. Perhaps there is no wave equation that can

25. You don't need to understand exactly how this works in order to continue our exploration, but if you'd like to see it, here goes. We want an equation like $E = Am + Bp$, but is also consistent with $E^2 = m^2c^4 + p^2c^4$. If we square the first equation we have to multiply all the terms by each other, so we get $E^2 = A^2m^2 + (Am \times Bp) + (Bp \times Am) + B^2p^2$. Now we have two expressions that ought to both equal E^2, so they must equal each other: $m^2c^4 + p^2c^4$ must equal $A^2m^2 + (Am \times Bp) + (Bp \times Am) + B^2p^2$, for any values of m and p. If we let A^2 and B^2 both be equal to c^2, that takes care of the terms involving the square of the mass and the square of the energy. But we are left with the other bit $(Am \times Bp) + (Bp \times Am)$. This can only be zero (for all values of p and m, because we want an equation that works for all particles) if $A \times B$ is equal to the negative of $B \times A$. But for all numbers, $A \times B$ equals $B \times A$. So A and B cannot be numbers.

properly describe electrons and at the same time agree with relativity. Maybe these waters are too rough for us, and the roads are too difficult to pass.

Not for Paul Dirac. He met the challenge—to find an equation that describes the motion of a quantum particle moving at relativistic speeds—in 1928. It involved a strange form of mathematics not used so far in this exploration, but luckily, waiting there in our tool kit to be used.

The important first step is to note that there are mathematical objects that do not commute. One such type of object is called a matrix. In mathematics, a matrix is an array of numbers arranged in rows and columns, governed by rules dictating how they should be multiplied together and so on.[26] Mathematicians do this kind of thing quite a lot—define some new abstract mathematical object with some specified behavior and then play around to see what the consequences are. From the point of view of the mathematics, it matters not at all whether there is any correspondence between the new toy and the physical world. But from the point of view of a physicist looking for some things that don't commute to help make more sense of the map of

26. From the point of view of mathematics, the numbers can be pretty much anything you like, although if we want to use them to help us understand physics, they take on specific meaning. For example, the numbers in a matrix can encode how the different components of a magnetic field change when we rotate it through some angle around some direction.

physics, it can be a treasure trove. Matrices can slot in there instead of numbers in the trial relativistic wave equation we are trying to build. They make it work, and then the mathematics of matrices can help us see what the physical predictions of the equation might be. In Dirac's case, they are dramatic.

The simplest matrices that can make the Dirac equation work are arranged in four columns and four rows, of four numbers each. These multiply the quantum field describing a particle, just like the terms in the Schrödinger equation. But the rules of the matrix game mean that the objects they multiply can also no longer be simply numbers. They have to have four components, arranged in a little column. Once you have decided you need matrices, you need four quantum fields to describe a particle, not just one. As one might expect, the fact that there are four components to the quantum field has real physical consequences. This will allow us to travel into the deep south, and also explain one of the mysteries we encountered long ago, back in Atom Land.

12

Spin and Antimatter

From the south coast of the island, the weather to the southeast looks grim and forbidding. The more superstitious crew members fear that we will be lost in the deep, or even off the edge of the world, if we set out in that direction. But with the Dirac equation, and its matrices and its vectors containing four quantum fields where we used to have one, we have constructed a vessel worthy of the journey, and it will take us on one of the most remarkable expeditions in the landscape of particle physics. One thing that gives us confidence in our new vessel is that, in developing it, we have solved an old puzzle we came across in Atom Land.

On that expedition we made the observation that the electrons bound to atoms were confined to particular energy levels. A curious fact that was also noted was that each level could only hold two electrons. There's no argument about this fact—the whole of chemistry and spectroscopy back it up. But why two? Even if we accept that there is some limited occupancy, that we can't stuff lots of electrons into the same energy level, two still seems like a strange number to allow. Why not just one? Why not ten?

The "exclusion principle" behind this is built deeply into the quantum field theory we use to describe electrons, and indeed all quantum particles, and as is very often the case, it comes down to a symmetry. In this case it is a symmetry under swapping pairs of identical particles.

That's a fairly trivial-sounding symmetry. One would think that swapping pairs of identical particles makes no difference to anything, by definition—they are identical. But in quantum mechanics that's not quite true. The probability of measuring a real physical property depends on the magnitude of the quantum field, not on its sign. Negative one and one both have the same magnitude—one! So swapping two identical particles will obviously have no effect on physical observables if it has no effect on the quantum field (that's trivial), but it will also have no effect even if it flips the sign of the field to negative whatever the original value was. Quantum particles seem to exploit all available opportunities to be weird, and some particles behave in the first way (no change in the field or the observable physics) whereas others behave in the second way (change the sign of the field, but still no change to the observable physics). The first kind of particles are called bosons, the second are fermions.

Now what is the chance of having two or more identical electrons in the same energy level in an atom? We have to, as usual in quantum mechanics, sum up all the different possible ways. Electrons are fermions, so there are two possible ways, according to this exchange symmetry—one with the particles swapped, one un-swapped. But these have identical quantum fields behind them, apart from a different sign. They will cancel out, just like the waves in the bay could arrive out of phase and leave the seagull becalmed. If they cancel out, the probability of having two identical electrons in the same energy level is zero.

That's the exclusion principle. It shows how the energy levels can get full, but it does not yet explain why there are *two* electrons in each energy level. According to the exclusion principle, there can be only one.

Unless . . . the electrons are not identical. And the Dirac equation has just shown us that the electron has to correspond to four quantum fields. Two of these explain the double occupancy of the energy levels in Atom Land. They correspond to different "spins" of the electron. Spin is an intrinsic angular momentum that electrons have, almost as though they were really spinning. It is a very important property of a particle, dictating a lot about how it behaves. Electrons have half a unit of spin, which can point in one of two directions—say, clockwise or counter-clockwise. The spin means electrons generate a tiny magnetic field, as well as the electric field due to their charge. This magnetic field affects how they bind to atomic nuclei. It can be measured—for example, if an atom is placed in a strong magnetic field, each energy level splits into two, because when the magnetic field due to the spin is aligned with the strong magnetic field, it has a different energy from the one it has when it is anti-aligned. These splits in energy levels have been precisely measured. The Dirac equation tells us why they happen.

The exclusion principle, and whether a particle is a boson or fermion, is in fact intimately wrapped up in the matter of spin. Fermions, such as the electron, have half-integer spin. Bosons, such as the photon, have integer spin.

But what of the other two components in the Dirac equation? This is where we launch our new boat and head south, into those storms.

At the time of Dirac's result, there was no physical property, like spin, waiting to be explained by these components. They look a bit like the negative energy solutions that plagued us earlier, but rather than negative energy, they describe particles that have the opposite charge to electrons, and identical mass.

No positively charged electrons were known. This was a problem for Dirac, and he wrongly suggested that the proton might be the so-called antiparticle of the electron, even though their masses are very different. He did not need to worry. Physicists were busily photographing the fragments produced when high-energy particles from space hit Earth's atmosphere. Within a very few years of Dirac's prediction, tracks were seen in these photographs that indicated, from the distance they traveled and the way they curved in a magnetic field, that they had the same mass as the electron, but were positively charged. They were named positrons—the antiparticle of the electron. As we have built better detectors and high-energy accelerators, antiparticles for all the other particles described by the Dirac equation have since been found.

This is what we find as we sail south. The world is a globe, not a flat plane. There is an equator, and sailing over it reveals a whole new landscape. The hemisphere south of the equator is antimatter, and the new vessel that takes us there is the Dirac equation, powered by mathematics, relativity, and quantum mechanics. Every type of particle we discover north of the equator has its southern, antimatter equivalent. Positrons for the electron, antiprotons for the proton, and so on.

Antimatter is one of the more astonishing concepts in the landscape of particle physics. In science fiction it appears as a weapon of awesome power, destructive or otherwise, largely because when matter and antimatter meet, they annihilate one another in a blaze of released energy. In real life it has, perhaps surprisingly, been put to use in medicine before warfare. It was a stunning prediction of new physics, pulled right out of human imagination and reason, shortly before experiment would reveal it to be correct. For all its starring roles in science fiction, antimatter is science fact. A whole new hemisphere of physics was made accessible by the exploratory power of some rather esoteric mathematics. The mathematics told us something startlingly new about nature, but perhaps an even more profound shock is what the observable existence of antimatter tells us about the connection between abstract mathematics and physical reality. The fact that the vessels we built have such range and power is astonishing.

To add the icing on the cake, just before we turn back toward the island on which Port Electron is found, another coastline drifts past in the murk. One of the more experienced sailors informs us that this is the land of Bosonia, and it is there that the photons we have come across so often on the road originate. This is clearly a country to be visited on a future trip. For now, we have business back near Port Electron again.

13

The Electron's Overweight Siblings

Dirac's equation, Maxwell's equations, and relativity are all brought together in QED, the quantum field theory we have already been using to navigate on land using our map. Back from the south and armed with our new knowledge, we now land yet again at Port Electron, and set off to explore the inland area around it, which we neglected earlier when we were intent on sailing for Atom Land.

We know quite a lot about the electron by now. We know it has a small mass. We know its connection to Atom Land, and the fact that it has a spin and an antiparticle. What else might be lying hitherto undiscovered in the hinterland of Port Electron? There are stories. The locals refer to the country as the Isle of Leptons. When asked what this means, they say that because it is so light, the electron is called a lepton, from the Greek for "small." But the question that poses itself then is—are there other leptons around?

A rapid trip on the excellent highways in our increasingly fast and roadworthy vehicle reveals that yes, there are. Two other particles that exist on this island are identical in every way to the electron except for the fact that they are more massive. The particles both carry electric charge like the electron, so they interact via the electromagnetic force—we can get to them by road, on our map. They also carry half a unit of spin; they are fermions. So far, so much like the electron. Their greater mass, however, is a significant difference, with major consequences.

Because they are massive, they are able to decay to lighter particles. Particles decay whenever they can, in general. Because mass is a form of energy, a particle is really a very, very dense clump of energy, which, if it can, will naturally tend to spread out, to distribute itself more evenly, by decaying to lighter particles. Imagine if I were to superheat a cubic centimeter of air in a room. That would be a concentration of heat energy, but it wouldn't last long. The hot air would spread out through the room almost immediately. The overall temperature would rise slightly, but rather quickly there would be no sign of where the hot cubic centimeter was. In a similar way, particles decay and spread out their energy. And the more massive they are, the more ways they have of doing this, because there are more options of lighter particles open to them.

The electron has no options. It is the lightest electrically charged particle. Electric charge is conserved, and there is nothing it can decay to that could carry the charge, so it does not decay.

Some way to the east on the Isle of Leptons, however, lies the muon. The muon is more than two hundred times more massive than an electron. That is a huge concentration of energy that will spread out very rapidly if it can. And there is, of course, a lighter particle that can carry away the charge of the muon—the electron itself. This is the only decay available to the muon (in the Standard Model, anyway). It decays to an electron, plus a neutrino and an antineutrino, which are very light and carry no electric charge. They occupy a different and very inaccessible part of the Isle of Leptons, and so far we have only heard rumors of them. The average time a muon exists before it decays this way is just over two microseconds. For this reason, muons play no significant role in Atom Land, although it is possible to form short-lived "muonic" atoms, in which one of the electrons bound to the nucleus is replaced by a muon. The most common source of muons is the upper atmosphere, where high-energy particles from space—cosmic rays—collide with the oxygen and nitrogen molecules of the planet's protective blanket.

Farther on down the road eastward is the tau lepton. This is heavier still, with a mass nearly seventeen times that of the muon. Because of this, it has more decay options open to it (always with at least one neutrino involved). The average time a tau stays with us before it decays is just one third of a millionth of a microsecond. This is just about long enough for the decays of fast-moving taus, produced in high-energy collisions in particle colliders, to be observed in particle detectors, but it is not long enough for them to form any bonds with atoms, or with anything else.

There is more to be learned by studying these new leptons more closely. As we know from QED, magnetic fields are created whenever electric charge moves around. So because of their spin and their electric charge, electrons, muons, and taus are tiny magnets. They have two poles, north and south (which together form a "dipole"), like Earth or any other magnet.

The strength of the magnetic dipole of a particle is an important quantity to measure. It depends on the spin, the electric charge, the mass, and a constant that is conventionally called *g*. The value of *g* tells us a lot about the particle in question.

For a classical, everyday particle, *g* is equal to negative one. Imagine a spinning ball with an electric charge spread equally throughout its volume. If we use Maxwell's equations to calculate the magnetic field due to that charge moving around the axis of rotation, we arrive at negative one. We always will.

However, electrons and muons are not classical, everyday particles. They are quantum particles, or fermions, described by the Dirac equation, as we have already seen. As part of the way the equation introduces a spin of half a unit, it also predicts that *g* should be negative two.

Compared to measurements of *g*, this is nearly right. The actual answer for the electron is –2.00231930436152, with an uncertainty of 0.00000000000054. That makes *g* one of the most precisely measured and calculated quantities in the world.

The value for the muon is very similar: –2.00233184178, with an uncertainty of about 0.0000000012. It is less precisely measured than the electron, as one might expect since muons are less common than electrons, and because they decay so quickly. But more interestingly, the theory and the most recent measurements disagree in the case of the muon, by about 3.4 standard deviations, which is expected only about 3-in-10,000 times if both theory and experiment are correct. That is enough of a discrepancy to motivate lots of effort both to calculate and to measure the value more precisely.

The reasons for these tiny differences from the Dirac prediction of two, and the reason for the interest in measuring this property of the particle so precisely, are the same: quantum corrections.

Several times on our travels we have come across the idea that, in quantum field theory, all possibilities have to be taken into account. This is how we get the right answer for the interference patterns when particles pass through slits. This is why electrons around atoms are confined to specific energy levels, and why each level can contain only two electrons. And we need to do this even for a particle that is just on its own, minding its own business.

Imagine that we pause by the roadside and watch a particle, say an electron or a muon. The particle is there one moment, and . . . continues to be there the next. The simplest possibility is that nothing happens. But there is also the possibility that in that moment, it briefly kicked off a photon, which sprang away and then came back again, so that at the end of the moment it is as though nothing happened. That possibility needs to be taken into account in calculating the properties of the particle, such as its magnetic dipole. There are even more esoteric possibilities. The photon might have split into a particle-antiparticle pair, which then annihilated with each other back to a photon, which was then reabsorbed by the original particle, so that again it is as though nothing happened. That possibility needs to be taken into account if you want a precise enough answer.

It is these tiny quantum loops and corrections that, ridiculous though they may sound, actually have an effect on *g*, the number that characterizes the magnetic moment, and are the reason it is not exactly two. The mass of the original particle also affects them, which is why *g* is slightly different for the electron and the muon.

The reason that knowing these values so precisely is interesting is that the tiny, transient loops might contain unknown particles not present in the Standard Model. Tantalizingly, this might be the reason that, for the muon, the measurement does not agree with the theory. Perhaps mysterious unknown particles not present in the Standard Model and not yet directly observed are going around those loops and affecting the value of *g*. If, with more precise measurements, the disagreement between the data and the Standard Model grows, this may be a definite signpost to what is going on in the far east of our map. Looking in detail at the inhabitants of the Isle of Leptons could have far-reaching consequences.

REST STOP

Gravity: A Distant Diversion

Interlude in a playground – Pretentious forces – The geodesic, straight lines that loop – Black hole mergers waving from a distance – A prediction vindicated

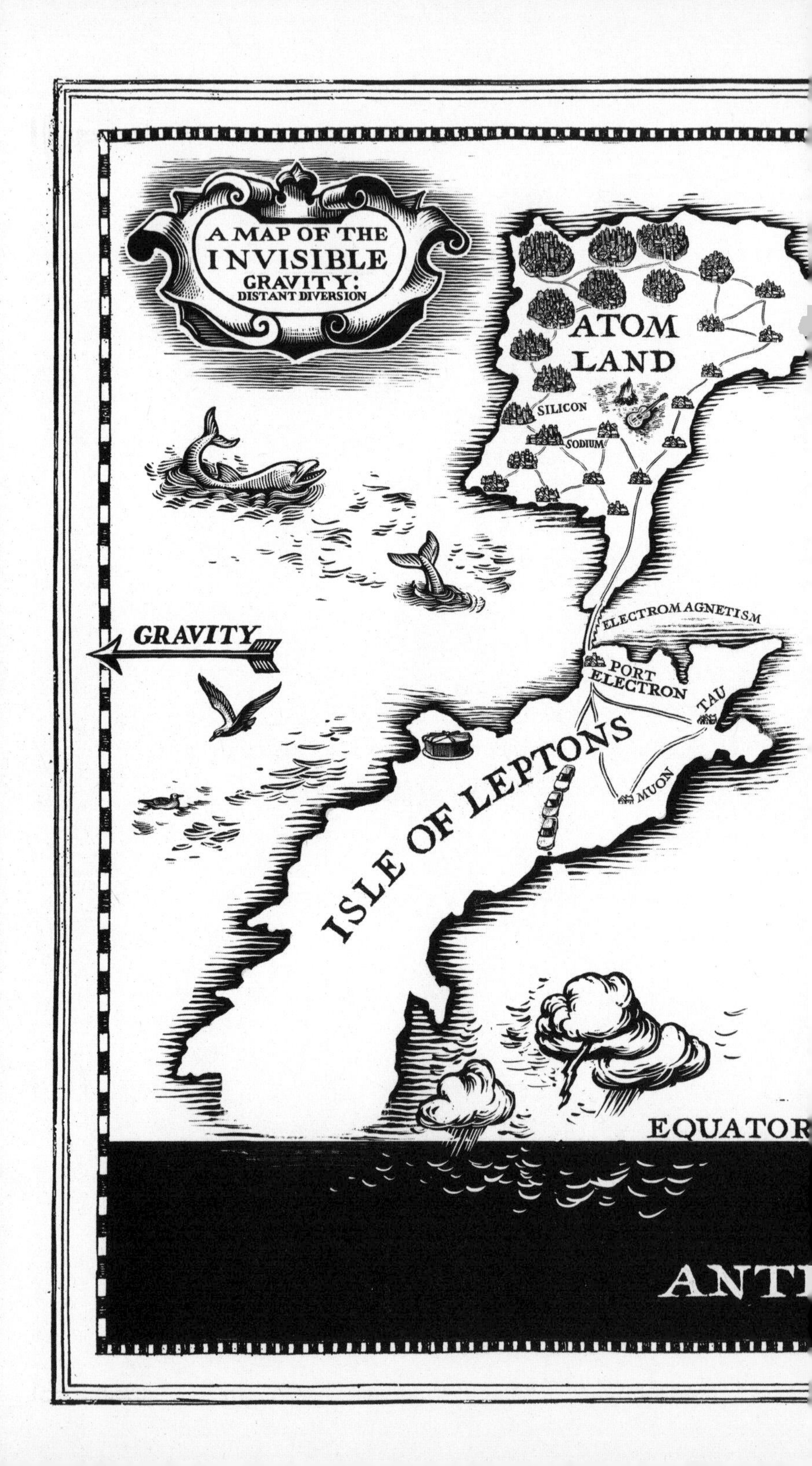

A MAP OF THE
INVISIBLE
GRAVITY:
DISTANT DIVERSION
ATOM
LAND
SILICON
SODIUM
GRAVITY
ELECTROMAGNETISM
PORT
ELECTRON
TAU
ISLE OF LEPTONS
MUON
EQUATOR
ANTI

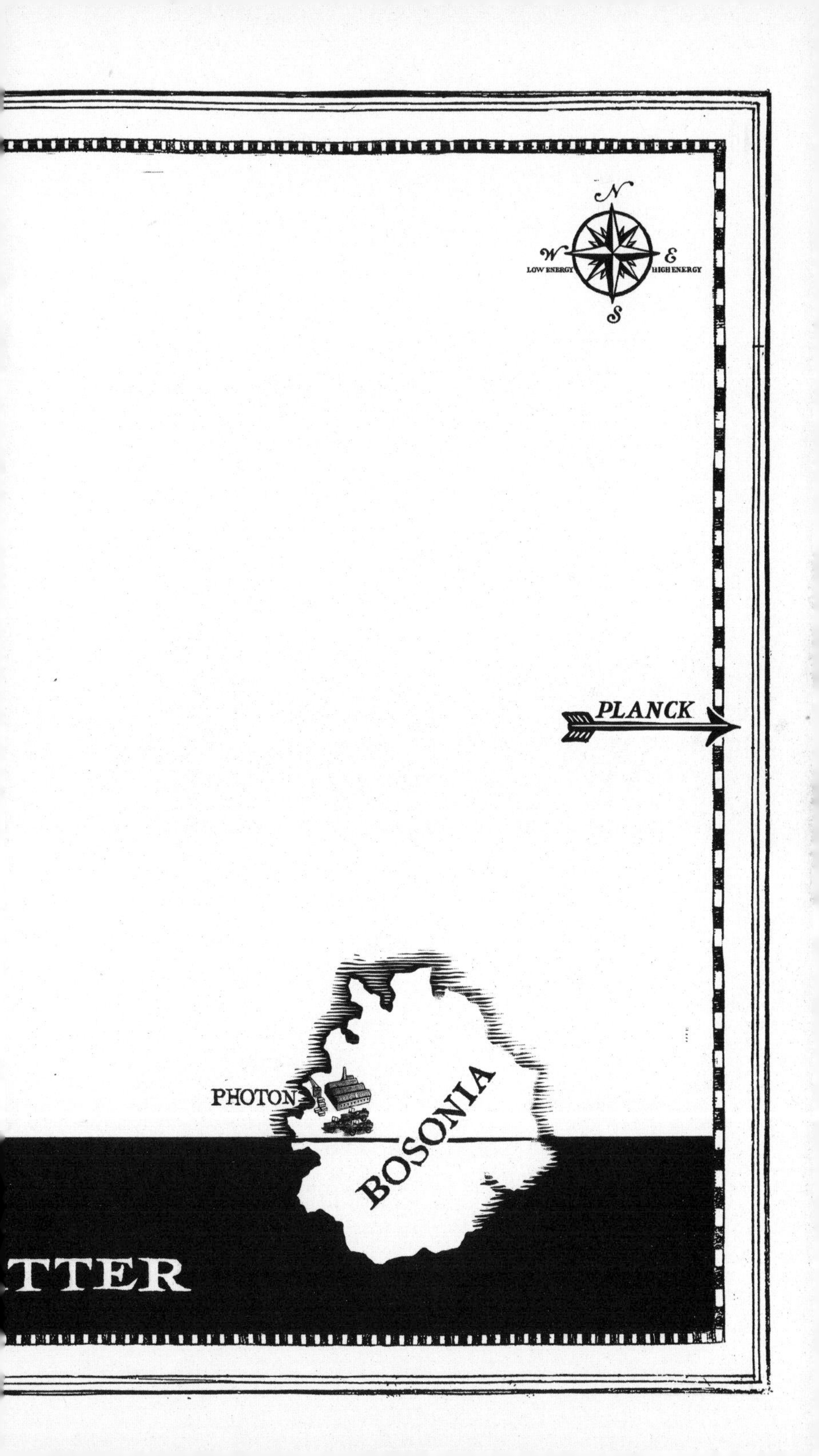
N
W
E
S
LOW ENERGY
HIGH ENERGY
PLANCK
PHOTON
BOSONIA
TTER

The Weakest Force

We have made a lot of progress in our explorations. Starting from Port Electron, we have surveyed the interior lives of atoms—discovering the nucleus and understanding the arrangement of the electrons that are bound to it—in our travels through Atom Land. We have found the muon and tau—two heavier copies of the electron—and explored the western end of the Isle of Leptons in detail. And we have worked out how the electromagnetic force operates, followed the road network, and even spotted the home of the photon on the west coast of Bosonia, although we did not land there.

There are more journeys to come, and there is more to be revealed. But there is one vital piece of physics that is operating in the background of all this, that is apparent in everyday life even to the far west of our map, and that we are getting no closer to understanding by traveling farther.

The force of gravity is felt by all matter and energy and is the most obvious force in everyday life. In fact, it is so ubiquitous, and such an accepted part of everyday life that—as the apocryphal story of Newton being inspired by a falling apple indicates—even noticing it as a force is something of a breakthrough. We get most excited when we appear to escape the force of gravity. An astronaut orbiting in free fall is a huge novelty when normal life is spent pressed to the ground inside a gravity well.

It is a bit of a surprise, then, to find that, of all the fundamental forces, gravity is the one with the most tenuous connection to our map of physics. There isn't really a "Gravity Island" or anything similar to explore. So we will sit ourselves down in a park somewhere in the suburbs of Port Electron on the Isle of Leptons, have a cup of tea, and think about it.

There are a few reasons for the distance between gravity and anything else on our map. The most obvious is the strength of the force. Gravity has a noticeable effect because we live next to an enormous mass (the Earth) that orbits an even bigger mass (the Sun). We are continually under the influence of the cumulative attraction of all the billions of trillions of atoms in those bodies. But those bodies exist far to the west of Atom Land, far off the map, too large and diffuse for its scope.

Thus, gravity is by far the strongest net force acting on us. Something to think about as we pour ourselves that cup of tea. And as we do so, and when we raise our cup to drink, our arm is countering the gravitational attraction of the entire Earth. In fact, we are overcoming gravity all the time, whenever we stand up, breathe, and just generally don't collapse into an unstructured jelly.[27] The force that allows us to do this is the electromagnetic force governing the chemical interactions in our bodies—the bonds that keep our bones rigid, the energy transfers that make our muscles contract and expand.

Unlike gravity, which is always attractive, electromagnetism can be both repulsive and attractive; there are both positive and negative charges. Since Earth and our bodies both contain equal amounts of positive charge (in the atomic nuclei) and negative charge (the electrons), the attractive and repulsive forces cancel each other out. Even a very small, local imbalance in this cancellation leads to significant effects. Lightning is a dramatic example, where an imbalance builds up in the atmosphere and balance is restored in a violent flash.

27. Most days except Monday, in fact.

This cancellation is why we don't feel electromagnetic forces as obviously as we feel the force of gravity. But down at the subatomic level, the electromagnetic force between a proton and an electron is about 10^{43} (one followed by forty-three zeros, or ten million billion billion billion billion) times stronger than the gravitational force. Considering this huge difference, particle physicists can ignore gravity in their calculations.[28]

28. Though, disappointingly, not in their personal lives.

Planes and Merry-Go-Rounds

The other way in which gravity is different from the forces of the Standard Model is related to the way it is understood within Einstein's theory of general relativity. In this theory, it is questionable whether we should even call gravity a force at all.

When getting to grips with the Dirac equation for our journeys across the map, we saw that the "special" theory of relativity arises when you insist that the laws of physics are the same for all "inertial observers"—observers traveling with constant speed in a constant direction. The laws of physics include electromagnetism, so that means that the speed of light must be the same for all those observers, and once you insist on that, all the weird effects on how time passes and how space contracts follow. As does the famous equivalence equation between energy and mass, $E = mc^2$. To get such astounding new physics from such a general principle is remarkable.

But what about other observers? Most of the time we are not "inertial observers." Surely the laws of physics should not change just because we are accelerating or decelerating? It is certainly tempting to insist that physics should be the same for all observers, even "non-inertial" ones, if only because we got such a lot of new and correct physical understanding from doing that for the inertial ones. Anyway, it seems intuitively right, somehow.

That is the challenge Einstein addressed with his general theory of relativity—"general" because it applies to all observers, not just to the special case of inertial ones considered in his previous masterpiece, the special theory of relativity.

What are the effects of acceleration? There is an airport not so far away from us—we will visit this soon on our travels. In an airplane accelerating down the runway for takeoff, the passengers feel themselves pressed back into their seats. From their point of view this feels like a force, and that's why they are not inertial observers. Imagine yourself in their place. The water bottle you left on the floor beneath your seat rolls backward as though someone is pushing it. Objects in your frame of reference do not move with constant speeds, as they should for an inertial observer. From your point of view, objects accelerate backward. From my point of view, watching from the park, you are accelerating forward and leaving them behind.

How does this connect to gravity? Well, imagine it is night. It is pitch-black outside. You are pressed back in your seat as the plane accelerates horizontally. But could you tell the difference between this situation and the possibility that the airplane might be flying upward at some angle and a constant speed?

In both cases the bottle (or carelessly stowed laptop) would accelerate backward until it hits something, maybe the back of the cabin. Einstein's key observation was to notice that it is difficult, perhaps impossible, to tell the difference between a frame that is non-inertial because it is experiencing a gravitational force (the climbing airplane) and a frame that is non-inertial because it is accelerating (the horizontal airplane speeding up).

Similarly, a frame that is falling freely under gravity looks very much like an inertial frame, in which no gravity acts. In fact, the only frames we're familiar with that really look inertial are those in free fall—most obviously, the frame moving along with the International

Space Station (ISS), which is continually free-falling along its orbit around Earth.

It is worth thinking about the situation of something in orbit like the ISS, compared to someone or something on, for example, the merry-go-round in the children's playground we've just noticed across the park from us.

From the point of view of someone on Earth, the ISS is moving quickly, and, but for the gravitational attraction between it and Earth, it would move in a straight line, according to the conservation of momentum. The gravitational force plays the same role as the arms of a child holding on to the merry-go-round, spinning in the playground we can see. By holding on, the child's arms exert a force that pulls them continually toward the center of the merry-go-round and so keeps them turning around it. Likewise, if gravity suddenly stopped working, the ISS would fly off into outer space. But gravity keeps it circling in orbit. Gravity provides a centripetal force, directed toward the center of Earth.

But there are some big differences between the situation of the child and the ISS, apart from the obvious ones involving spacesuits and stuff.

On the merry-go-round, the child experiences a centrifugal "pseudoforce." They definitely are not in an inertial frame. Anything they drop will fly off the merry-go-round, away from the axis of rotation in the middle. Yet on the ISS, the astronauts seem to experience fewer forces acting. They not only do not experience a centrifugal pseudoforce, but also seem to be weightless.

The reason is that in orbit, your weight and the centrifugal pseudoforce completely cancel out, leaving you in free fall. This is a very different experience from being flung around a merry-go-round, for two main reasons.

Firstly, gravity acts on your whole body, and indeed all matter, equally and simultaneously, so you don't have to grab onto a handle and then have your arms pull the rest of you around the circle. Your arms, legs, and the rest are all being acted on by gravity.

The second reason goes to the heart of general relativity and the reason you are, on the ISS, in fact in an inertial frame.

Mass appears in two important equations—force is mass multiplied by acceleration,[29] and the force of gravity is proportional to mass. General relativity works because the mass in those two relations is identical. On the face of it there is no reason that this should be so, but in general relativity it is built in. That was another of Einstein's great insights.

This means that in your ISS reference frame, the centrifugal pseudoforce can be canceled out by the force of gravity, not just for one particular mass

29. Newton's law, $F = ma$.

approximately, but for *all* masses, exactly, *all the time*, leaving you in an inertial frame. In fact, because it cancels a pseudoforce in this way, it is legitimate to say that the general relativity reduces the "force of gravity" to the status of just another pseudoforce.

The principle of conservation of momentum, used to define an inertial frame, still applies and still defines such a frame, but inertial frames now include any frame falling freely under gravity. Bodies falling freely like this are said to be traveling along a "geodesic," which is a redefinition in space and time of what constitutes a straight line, the shortest route between two points.

In the absence of a gravitational field, a geodesic is a straight line in the Euclidian sense. Euclid was the first person we know of to define rules for geometry. Among these rules, or axioms, is the statement that the maximum number of times a pair of straight lines can cross each other is once, and parallel straight lines never meet each other. If there is no gravitational field around, geodesics are straight lines, and general relativity, special relativity, and Newton's laws of motion all agree that freely moving bodies will travel along a straight line at a constant speed.

But near a large mass, general relativity states that geodesics bend into curves, or even into the closed ellipse of an orbiting ISS. Space and time, defined by geodesics, are no longer Euclidian. The very meaning of a "straight line" has changed.

Maybe the easiest way to get an idea of what is going on is to imagine two people at the equator a few miles apart, setting off due north, parallel to each other. Though they set off parallel and keep traveling in the same direction—due north—they will eventually meet, at the North Pole, because the surface of the Earth is curved and defines a non-Euclidian two-dimensional geometry.

Near a large mass, space is curved in three dimensions, and "straight lines"—geodesics—can become orbits. This curvature is what we, and all other masses, experience as gravitational force. This is why gravity is, in some sense, a pseudoforce. It is an effect generated by curves in the geometry of space-time.

This is a way of thinking about a "force" that is very different from how we think about the forces in the Standard Model. If we think of particles and forces as actors on the stage of space-time, gravity isn't just another actor; it is something that bends the stage on which the others perform. Also, as is already being discovered in our exploration, the other forces are fully developed in quantum field theories, while gravity is definitely not.

Different, yet Somehow the Same

There are some things in common between gravity and the other forces. The ideas of symmetry play an important role in both. Any time we change something (say, rotate a shape) and it makes no difference (if the shape we rotated was a sphere), there is a symmetry. In quantum mechanics, there is a symmetry in that changing the phase of all quantum waves at the same time makes no difference to the physics. In general relativity, moving around between different accelerating or moving frames, in or out of gravitational fields, makes no difference to the physics. A lot of theoretical effort has gone into trying to exploit such similarities, with some very powerful results, but so far no one has managed to make a fully Standard Model–like quantum theory of gravity that would work at very small distances and high energies.

Stepping back from the quantum world, however, there is a very simple similarity between gravity and electromagnetism.

The Sun pulls Earth toward it with gravitational force. If Earth were twice as far away from the Sun, the force would be four times weaker. If Earth were three times nearer to the Sun, the force would be nine times stronger. This is the famous "inverse square" law. Multiply the distance by two, and the force gets weaker by the square of two—that is, by a factor of four. Shrink the distance to a third, and the force gets stronger by the square of three, i.e., nine.

It tickles me that even though the theories behind the forces are so very different, the electric force does exactly the same thing. There is an attraction between a negatively charged electron and the proton in a hydrogen atom. Double the distance between them and the force drops by four; it follows an inverse square law, just like gravity.

This is no coincidence. We can think of a mass, or an electric charge, as the source of a force. Physicists often draw them as lines of force.[30] The density of the

30. An innovation due to Michael Faraday's very early thoughts about electricity and magnetism.

lines is proportional to the strength of the force. The force is spread out over a bigger and bigger sphere as you move farther and farther from the source. If the total number of force lines stays the same, then the force at any given point will drop as the area increases. The area of a sphere is four times π times the square of the radius of the sphere, so the amount of force at any given point on the sphere is divided by this. The most important thing in that expression is the radius. The force is divided by the radius squared. This is the "inverse square law," which works just as well for any long-range force, whether the theory behind it is based on quantum mechanics or warped space-time.

This is a good example of the fact that some of the general features that we encounter on our travels across the map are in a sense more fundamental than the details. The principles that emerge are in some ways more fundamental than the underlying theory. Conservation laws and symmetries are other examples of the same effect. You don't need to know all the internal details of water molecules to have a good idea of what will happen if you boil a kettle. You don't need to understand QED or general relativity to know that the inverse square law is probably a good bet to describe how the force falls off with distance.

Ripples in the Space-Time Continuum

Continuing the attempt to find similarities between gravity and electromagnetism, the first real theory of electromagnetism, encapsulated in Maxwell's equations, is not a quantum theory, either. It describes the electromagnetic interactions in terms of continuous fields, just as general relativity talks about continuous space-time. And Maxwell's equations predict the existence of electromagnetic waves—light, X-rays, radio, the whole spectrum—before we even start discussing quantum effects. Does general relativity do something similar? Are there waves in gravitational fields?

As far as the theory goes, the answer is decisively yes. After some early uncertainty, Einstein and everyone else who looked into it agreed that gravitational waves should exist.

They have to exist in some sense, because a moving mass will cause changes in the gravitational field—changes in the curvature of space-time—in much the same way that the dolphins we saw earlier caused ripples in the bay. According to relativistic principles, those changes cannot appear instantaneously all over space. They can only spread around at the speed of light, and that implies a ripple, a wave of curvature, traveling away from the moving mass.

The waves warp distances in space, shortening them in one direction and lengthening others perpendicular to them, the way that, if you squeeze the circular rim of the plastic cup containing your tea and let go, the rim is squashed into an ellipse first one way, then back the other way.

Until September 2015, gravitational waves were expected but never seen. To see them would require something much more cataclysmic than squeezing a plastic teacup. It would need to be a huge astrophysical event—stars or black holes smashing together or exploding. Even with such a huge event, it would require an unimaginably sensitive detector, simply because the force of gravity is so weak.

According to general relativity, as two objects orbit each other under their mutual gravitational attraction, they radiate energy as gravitational waves. Losing energy incredibly slowly, they will spiral very, very gradually inward, with their rates of rotation speeding up as they go, until eventually, after a super-rapid whiz around each other, they collide and either merge or smash each other to pieces. While the end is rapid and violent, the change in orbit early on in this process is so tiny that even for enormous stars or black holes, the waves are too faint to detect.

There was evidence that they were there, though. In 1974, Russell Hulse and Joseph Taylor Jr. at the University of Massachusetts published precise measurements of the first "binary pulsar." A pulsar is a stellar object that emits regular pulses of electromagnetic radiation. In the Hulse-Taylor pulsar, the pulses show a shift in their radio frequency. When the source is moving toward us, the radio waves have a higher frequency, and when it moves away they have a lower frequency. The two orbiting bodies in that binary pulsar are ultra-dense, with radii of about ten kilometers but masses comparable to that of the Sun. They are orbiting each other at a distance of a few times the distance between Earth and the Moon—practically on top of each other in astronomical terms. And they complete an orbit in under eight hours, so they are traveling very quickly.

Even this gargantuan, high-speed cosmic orbital dance does not make big enough gravitational waves for them to be directly observed. However, what could be observed, from the regular shift in the frequencies, was that their rate of rotation is increasing very slowly. This is happening at a rate precisely consistent with the rate expected in general relativity, due to the radiation of energy as gravitational waves.

This was a huge boost to the credibility of the idea that gravitational waves should exist. But it is still not the same as actually measuring them, and it tells you nothing about how they travel.

The beautiful elegance of general relativity is a little deceptive—actually solving Einstein's equation to get a real prediction, which tells you what kind of experiment you need to build to test the prediction, is a major mathematical challenge and a towering achievement in itself.

As gravitational waves pass through Earth they distort distances, compressing them in one direction while stretching them in the perpendicular direction like the rim of the cup. The amounts of compression and stretching are tiny, though; less than the diameter of a proton over several kilometers. It is astonishing that we can even dream of measuring such minuscule effects. But people have done so.

The key is waves again, and specifically the interference effects that we saw in the bay. A beam of light can be split, and the two halves made to travel down different paths, then brought back together again. If the two paths are the same length, or differ by a whole number of wavelengths of the light, the two halves of the beam will be in phase—the peaks will line up with each other, as will the troughs. But if there is a difference in the paths of half a wavelength, they will be out of phase and will cancel each other out. The observed intensity of the light is very sensitive to fractions of a wavelength. An instrument that uses this sensitivity to measure distance is called an interferometer.

For optical or infrared light, distances of a few hundred nanometers (billionths of a meter) could be measured. Small, but nowhere near sensitive enough to see gravitational waves. However, if the two paths the light takes are long enough, and the light is reflected back and forth many times by mirrors, the sensitivity can be increased.

The Laser Interferometer Gravitational-Wave Observatory (LIGO) consists of two enormous interferometers, one in Hanford, in Washington state, and the other in Livingston, Louisiana. Each has two perpendicular arms two and a half miles long. An infrared laser beam is split, and half is sent up each arm. The beam is reflected more than 200 times, and the sensitivity is designed to be one-ten thousandth of the width of the proton, or a ten billionth of a nanometer. The technology required—accurate, efficient mirrors, powerful, stable laser, and so on—is formidable, but the principle is simple.

In February 2016, LIGO announced that they had seen gravitational waves consistent with the merger of two black holes. Further observations have been reported since.

This is a big triumph for general relativity. Everything we can see about the waves is as predicted by the theory, and they are now being used as a new means of observing astrophysical phenomena. They are likely to tell us a lot about the universe around us over the next few years.

In summary, much is known about gravity. General relativity makes precise predictions, correctly matching the motion of the planets and the fall of an apple, and the existence of gravitational waves. These waves are the analogue of radio waves for electromagnetism—described by Maxwell's equations before quantum electrodynamics came along. If there is a quantum particle carrying gravity—a graviton, the cousin of the photon that carries electromagnetism—these waves are what it looks like in the low-energy, classical limit. They tell us the graviton is massless, if it exists as a quantum particle. But they do not give us a quantum theory of gravity; they do not allow us to fit it into our map of particle physics. And as we will see as we travel much farther east, there are other reasons to worry about gravity, as well.

Speaking of traveling, though, our picnic in the park has been long enough. The crew is eager to be off, and we have business with nearer land masses. It is time to finish our tea, stop the speculation, and head off on our travels once more.

EXPEDITION IV

Great Train Journeys

Into the nucleus – Elements and their isotopes – A profusion of hadrons and the meaning of the eightfold way – An important crossing to a different kind of country – Various flavors – Following the gluon

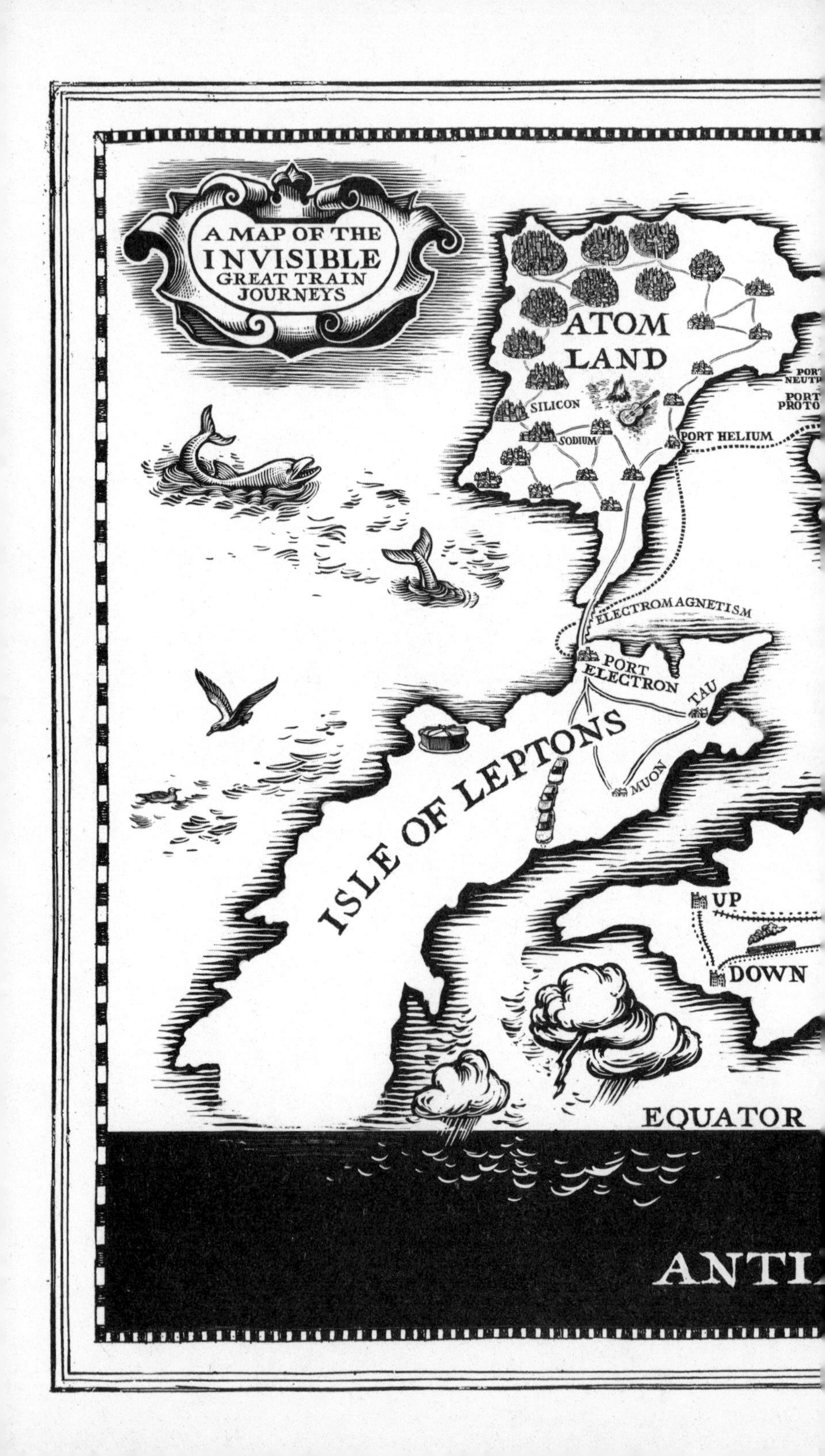
A MAP OF THE
INVISIBLE
GREAT TRAIN
JOURNEYS
ATOM
LAND
SILICON
SODIUM
PORT HELIUM
ELECTROMAGNETISM
PORT
ELECTRON
TAU
MUON
ISLE OF LEPTONS
UP
DOWN
EQUATOR

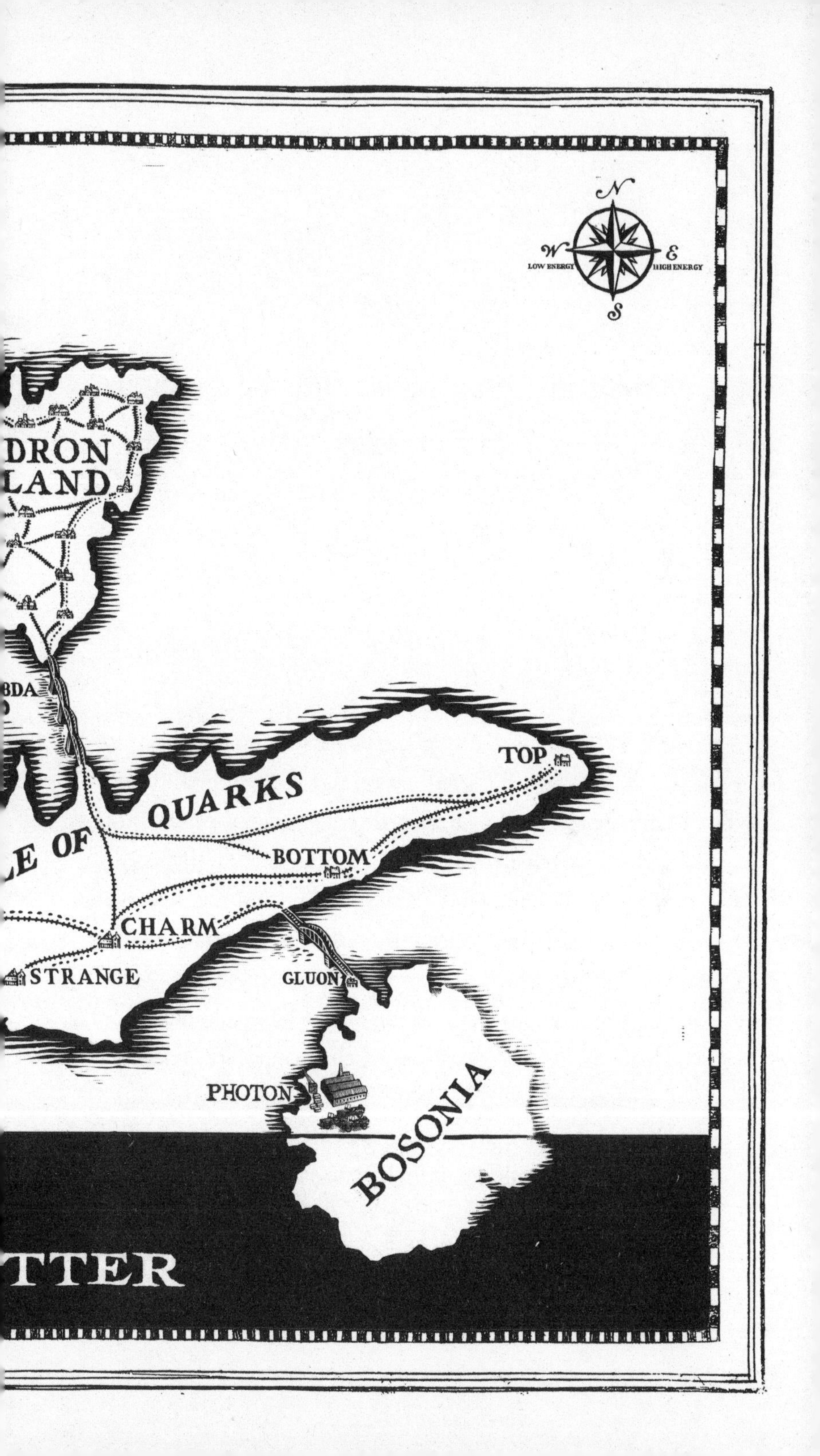
N
W
E
S
LOW ENERGY
HIGH ENERGY
DRON
LAND
BDA
LE OF
QUARKS
TOP
BOTTOM
CHARM
STRANGE
GLUON
PHOTON
BOSONIA
TTER

14

Protons, Neutrons, and the Nucleus

We are heading eastward again. That means toward higher energies, and smaller and more massive particles. Our last discovery was the tau lepton, which we visited by road, in the east of the Isle of Leptons. The tau has a mass of about 1.8 giga-electron volts, nearly twice the mass of a hydrogen atom. Looking northward from the vantage point of the tau, a new coastline looms through the mist, a hitherto unexplored land mass. The coastline of this same land mass was also vaguely visible from the east coast of Atom Land, where we discovered the atomic nucleus and toyed briefly with alpha particles. We return to our boat, departing once more from Port Electron, but this time sailing east. Our next journey will take us inside the atomic nucleus, and beyond.

During our exploration of Atom Land, the atomic nucleus was revealed inside the atom by a natural beam of particles—alpha particles—produced in the decay of the radioactive element radon. To look inside the nucleus itself, artificially produced beams of particles accelerated in electric and magnetic fields are required. We know from our studies of quantum fields that high-energy particles have correspondingly short wavelengths, and that is exactly what we need to resolve the internal structure of the nucleus.[31]

Such techniques reveal that alpha particles themselves have internal structure. They are actually two protons and two neutrons bound together, a particularly tightly bound, stable configuration. It turns out that all nuclei are made of some number of protons and some number of neutrons, with the single exception of hydrogen, the lightest element, whose nucleus is a lonely proton with no neutrons.

The alpha particle—two protons, two neutrons—is the nucleus of a helium atom. Helium is so light that it easily escapes from Earth's atmosphere and out into space. The only reason helium is still around in significant quantities is because it is constantly produced in the radioactive decay of heavy elements, such as the

31. Or to put it another way, they can smash it up.

radon used by Rutherford and his team. Your party balloons are full of alpha particles (with electrons stuck to them).

Nuclei of different elements are distinguished from each other by the number of protons they contain. We walked around silicon when exploring Atom Land and saw that the silicon atom is the smallest unit of silicon. The same is true for all elements. Although an atom of any element can be broken into still smaller pieces—electrons, protons, and neutrons—at that stage it will no longer be an element.

The silicon nucleus consists of fourteen protons and some neutrons. Anything with a different number of protons is not silicon. Thirteen would be aluminium and fifteen would be phosphorus, both of which are very, very different in their behaviors.

The history of studying the nucleus emphasizes an important fact about the map we are drawing of the fundamental structure of matter. As mentioned, like most scientific knowledge, the Standard Model is in some ways provisional. What we now view as fundamental particles, the smallest things in existence, may simply be the smallest things we can measure so far with the tools available. There may be deeper layers.

This is not to say the theory is not "true" in some sense, just that it might not be the whole truth, and a quote about the nucleus from Max Planck, the great

German physicist, one of the pioneers of quantum mechanics, illustrates the point. In his book *The Universe in the Light of Modern Physics* (1931),[32] he confidently explains uranium using the preferred model of the time:

> Uranium contains 238 protons and 238 electrons; but only 92 electrons revolve round the nucleus while the others are fixed in it. . . . The chemical properties of an element depend not on the total number of its protons or electrons, but on the number of revolving electrons, which yield the atomic number of the element.

There is an interesting mixture of wrong and right in here. Planck did not know about the neutron (discovered by James Chadwick the year after this translation appeared), so his description of the nucleus is wrong. Uranium 238 has only 92 protons, matching the 92 "revolving" electrons, but it has 146 neutrons nestled in the nucleus with the protons. We know that there are no electrons in the nucleus, in the sense Planck meant. We also know that the electrons aren't "revolving" in the same way planets orbit the Sun. This was an early model of the atom, proposed by Niels Bohr. As

32. A translation of the original *Das Weltbild der neuen Physik* (1929), G. Allen & Unwin, 1931.

we have seen already on our exploration, electrons are fundamentally quantum objects, not particles in the way Planck would have understood them at that time. However, he was absolutely right that there is a nucleus, with ninety-two electrons outside it. He was also right that these ninety-two electrons determine the chemical properties of the element. His model is useful, and would still work fine today as far as the chemistry of elements goes. But new data, from higher-energy experiments, surveying the area farther east in our map than physics had gone in Planck's day, would show that some aspects of it were wrong and needed to be changed.

This is another reminder that the Standard Model may need to be changed when new data comes along. For example, our confidence in the idea that electrons are infinitely small objects is determined by our current technology and the ingenuity with which we can use it to probe them. A new experiment may one day show that the electron is not an infinitesimal point, and contains still smaller constituents. The Standard Model would then need to be changed. In terms of our map, any changes would most likely extend eastward and reveal new features, although most of what we are exploring now would remain recognizable.

Changing the number of protons in a nucleus, then, changes the number of electrons the atom will attract, and so changes the chemistry. It changes the element.

Changing the number of neutrons has a less dramatic impact, however.

A silicon nucleus *usually* contains fourteen neutrons, but sometimes it has fifteen or sixteen. All three versions have the same chemical properties. They are all silicon. Different versions of the same element that differ only in the number of neutrons inside the nucleus are called "isotopes." Some elements have many isotopes, and sometimes they are unstable—that is, they decay radioactively to other elements.

Although the neutrons are not as crucial as protons in defining the behavior of an atom, they are absolutely essential in holding it together. Protons all have positive electric charges, and since like charges repel each other, any pair of protons up against each other in the tiny confines of the atomic nucleus must be experiencing huge repulsive forces. We should expect nuclei to fly apart because of this—or in fact never to form in the first place. The reason they don't is that they are attracted to each other by a force strong enough to overpower the electromagnetic repulsion.

But what is this force? It is a connection between features of the landscape that we have not previously encountered. We land at the Port Proton, on the shore of this new eastern island, the target of this expedition. We are granted access because of our high-energy particle beams that are powering us farther east. As we moor up at the strange, foreign harbor, we look for a road to follow inland. But a whistle is heard from inside a large building. We approach to investigate and notice a puff of smoke drifting out behind the building. Rails and sleeper cars emerge from the back, heading off into the interior of this new land. A sign above the rather grand entrance says HADRON ISLAND RAILWAYS: PROTON CENTRAL. We have found the strong force, the best way to explore Hadron Island.

15

Hadrons

Port Proton and the nearby Port Neutron are the first two examples of a type of particle called a hadron. Hadrons are densely scattered throughout the landscape we are now traveling by train. In nature, protons and neutrons are the most common hadrons; we already saw that they make up all the atomic nuclei, and so it is fitting that they are the first, most obvious things we come across on Hadron Island.

This is a land that is home to a complex network of structures tightly connected by railway lines. This is the strong force. The strong force is the second fundamental force of the Standard Model that we have encountered.

There is much to learn about it, but the most obvious fact right now is that it overcomes the electromagnetic force and holds the nucleus together. The way this manifests itself, as with any attractive force, is that the total energy of a collection of protons and neutrons—let's say two of each—is reduced when they are stuck together.

This is also the case for electrons bound to a nucleus; they stay bound because they have less energy that way. To release them, energy needs to be added. The main difference with protons and neutrons bound in nuclei is that the binding force is enormously stronger, and so the energies involved are much higher. We are, after all, moving farther east.

The large binding energies involved in holding nuclei together are seen in the values of the masses of the nuclei. Since energy and mass are equivalent, a lower-energy system has a lower mass. So the mass of a helium nucleus, for example, is less than two times the mass of a neutron plus two times the mass of a proton.

Again, this is also true of any bound system: The mass of a helium atom is less than the mass of a helium nucleus plus twice the mass of the electron, because of the electromagnetic binding energy. But the electromagnetic mass difference is tiny compared even to the mass of the electron. The mass differences due to nuclear binding energies are far more significant than this. This is why the energy that can be released in nuclear reactions is much greater than even the most energetic chemical reaction.

The binding energy per proton or neutron for different nuclei is biggest for elements around iron (twenty-six protons and around thirty neutrons). It drops rapidly for elements lighter than iron, which is why energy can be released by fusing them together, as happens, for example, in stars and hydrogen bombs and hopefully—if we can manage to compress and contain the enormous energies and densities required—in power stations one day. The energy also drops, although more slowly, as the mass of elements increases above iron. This means that there is energy to be gained not just by fusing light elements together to bring their mass up toward that of iron, but also by breaking down elements that are heavier than iron to bring their mass down toward it. This is nuclear fission, already useful as a terrestrial power source. In both fusion and fission, everything heads toward iron, which explains why it is such a common element, making up the core of our planet, for example.

The binding energy has another important impact. A neutron left on its own is unstable. On average, a free neutron outside an atomic nucleus and not bound to any protons will decay after about a quarter of an hour. However, once bound in a nucleus, a neutron is protected by the binding energy. That is, if it were to decay, the resulting fragments would have a higher total energy than the nucleus as a whole. Since decays must conserve energy, this is not possible, and the decay does not happen. This is why, while there are plenty of single protons around in the universe, there are no single neutrons, even though they exist in abundance in partnership with protons, inside nuclei.

All other known hadrons decay even more rapidly than neutrons. Most of them last for only tiny fractions of a second. This is why they were discovered later—their fleeting presence was first seen in the debris when high-energy particles from space bombarded Earth's atmosphere. These days they are copiously produced in collisions at high-energy particle accelerators.

As we traverse Hadron Island by train, it becomes clear that there are *a lot* of hadrons. They decay at different rates and in different ways, have different spins, and have masses ranging from the lightest—the pion, at about a sixth of the mass of the proton—up to several times the proton's mass.

We can also note that hadrons come in two types, baryons and mesons, with baryons being generally heavier than mesons.[33] The proton and the neutron are

33. As with leptons, this is from the Greek *barus* (heavy) and *meso* (intermediate).

examples of baryons, but there are many more, and dozens of mesons connected as whistle-stops or major stations on the line. This plethora of particles is disturbing if one is hoping for a simple set of constituents for matter. If all the hadrons were "fundamental," that would be a confusing array of building blocks.

However, as we travel the island, it is apparent that there are patterns in the arrangement of hadrons. Much as the periodic table arranges elements according to their reactivity, the mesons and baryons can be arranged in groups of eight or twelve (octets and decuplets) based on their spin, charge, and other properties. We see this as we go down the line: equidistant stations clustered together at a similar longitude and differing from each other in systematic, predictable ways. For example, we pass three almost identical hadrons, each differing by one unit of their electric charge, clustered with another five similar ones differing by one unit of spin. This kind of pattern is repeated across the length and breadth of Hadron Island.

The regularity of the periodic table is a big clue as to the internal structure of atoms. As we saw, the chemical properties of the atoms are driven by how many electrons they have, and how tightly bound they are to the nucleus. Similarly, the octets and decuplets of Hadron Island betray the internal composition of hadrons.

16

Quarks and the Strong Force

The analysis of the patterns of hadrons we have noted was systematized in the whimsically named "eightfold way" of US physicist Murray Gell-Mann, who introduced the concept of "quarks." All hadrons are made from quarks—baryons contain three quarks, and mesons contain one quark and one antiquark.[34]

The properties of hadrons come from the quarks they contain and the way they are bound together by the strong force. When Gell-Mann, and, independently, the Russian-American physicist George Zweig, first

34. Very recently, new structures have been found deep in the interior of the island that have two quarks and two antiquarks (tetraquarks) and four quarks and an antiquark (pentaquarks). Extremely ephemeral, the internal structure of these objects is the subject of ongoing exploration.

introduced the idea of quarks, all the known hadrons could be arranged according to the octet and decuplet patterns, and where there were gaps in the patterns, the expected hadrons were later identified experimentally. The position of a hadron in the eightfold way simply arises from which quarks it contains and in which direction they spin. Many of the fastest-decaying hadrons do so because the quarks inside them fly apart in different, more stable, combinations.

All of that summarizes some of the best evidence for the existence of quarks—they can be used to predict the existence of new hadrons with a particular charge, spin, mass, and so on. It has become clear that Hadron Island is strongly linked to another land mass to the southeast. The Isle of Quarks is our next destination. And we will travel by train—the strong force takes us directly there, over an imposing railway bridge.

Just as the electromagnetic force is carried by photons from their point of origin in Bosonia to travel the road network far and wide, the strong force is also carried by a boson. This boson, the trains on our tracks, is called the gluon—it glues the quarks together inside hadrons, and also sticks the protons and neutrons together inside nuclei. It is this force that overcomes the electromagnetic repulsion between the protons.

The strong-force equivalent of electric charge is called "color" charge, always using the US spelling, since it has nothing to do with optical color and was named in America. It was named this way because there are three "colors" of quark that when mixed together leave no net color, just as red, green, and blue mix together to form light. Quarks have color charge, and so do gluons. Analogous to QED in electromagnetism, this theory of the strong force is called quantum chromodynamics (QCD), and it is with QCD that all the calculations and predictions for hadronic properties and for most of the behavior of quarks and gluons have to be made.

Using QCD, it is possible to calculate important features of the strong force, such as the way it changes with distance. It is also possible to get information about the masses of the hadrons. Those masses are mostly due to the binding energy of the quarks and gluons, and the fact that they are whizzing around so rapidly inside the hadrons.

The strong force is not only strong, but also does not get weaker with increasing distance. The electromagnetic force and gravity both fall away as the square of the distance—so if two objects are moved farther away by a factor of two, the gravitational or electromagnetic force will fall by a factor of four. Not so for the strong force.

As two quarks are pulled apart, the strong force between them remains constant. This means a huge amount of potential energy builds up in the gap, and at some point there is enough of it to produce new quark-antiquark pairs, as if from nowhere. The quantum vacuum contains little loops of particle-antiparticle pairs, similar to those we saw affecting the magnetic moment of the electron. Making such a pair into a real pair of a quark and an antiquark can mean that the strong force has to stretch over a shorter distance, and that reduces the potential energy. So, because quantum mechanics allows it and the result lowers the tension, these pairs will indeed be spontaneously created. This is weird, and important, because these new particles then stick to the original quarks, making new hadrons. In other words, you have to put in so much energy to separate quarks from each other that you end up making more quarks. That is frustrating if your goal is to isolate a quark. Many experiments have searched in vain for isolated quarks.

However, as two quarks get closer, the strength of the force between them actually drops. This is what happens as we move eastward across Hadron Island, upward in energy scale—remember, short distance means higher energy. Atom Land was characterized by binding energies of several thousand electron volts. The nuclear binding energies on the west coast of Hadron Island are thousands of times higher, many millions of electron volts. Traveling eastward among the hadrons, we reached an energy scale of great importance, at around 200 million electron volts. This was where we crossed the bridge between Hadron Island and the Isle of Quarks.

Around this energy scale, which is known to the locals by the strange name of "Lambda QCD," a change in physics takes place. On the northwestern side, quarks are effectively invisible, hidden inside hadrons because of the weird pair-creation behavior as you try to pull them apart. But to the southeast, the energy level has increased to such an extent that we start being able to resolve the quarks inside the hadrons, to see them in their natural habitat. And what we see is that they are moving around quite freely, because they are very close together and the force between them is relatively weak.

As we travel east and the energy scale increases well above Lambda QCD, the strength of the force continues to decrease, and we can begin to explore the Isle of Quarks in ever more detail.

17

Life Beyond the Bridge

The Isle of Quarks is a confusing place. Since quarks can never be isolated, they never directly register in a detector, so it is reasonable to ask how we know they are there at all. The patterns we saw among the hadrons, described by the eightfold way, are compelling, but a bit more evidence is required. Are the units used in the eightfold way to arrange those patterns actual physical particles, or just a convenient mathematical trick?

As we travel into the Isle of Quarks, there are two important and direct ways in which the existence and behavior of quarks and gluons can be established. The first of these is the production, in high-energy particle collisions, of dramatic features known as hadronic "jets." They are sprays of hadrons, all traveling together in roughly the same direction, produced quite often when the colliding beams have an energy that places them east of Lambda QCD.

Jets are seen, for example, when electrons and positrons collide and annihilate at high enough energy, producing quarks and antiquarks. They are understood in QCD—the theory of the strong force—as follows: Pairs of quarks and antiquarks fly apart from each other, carrying all the kinetic energy of the initial colliding electron and positron. At first, they are very close together, and the QCD interaction between them is quite weak. The quark and antiquark will behave as though they are free particles. But very soon the tug of the strong force will be felt. As is the case whenever quarks try to escape, the potential energy between them increases as they separate, and this leads to the creation of more quarks and antiquarks. This process can happen several times, each time sucking away some of the energy of the initial quarks and producing more quark and antiquark pairs. In the end there are two sprays, or jets, of many quarks and antiquarks, which are heading off away from each other, going in roughly the directions of the initial quark and antiquark. Within these jets, the quarks are traveling close enough to each other that they can bind together and form hadrons.

The end result is usually two jets of hadrons, one for the quark and one for the antiquark. The sprays, or jets, of hadrons will be aligned roughly in the direction of the initial quark and antiquark. The energies and directions of the initial quark and antiquark can be calculated in QCD, and the calculation agrees well with measurements of the jets. This agreement is strong evidence that quarks really do exist. Good to know, as the stations fly by.

Jets also provide evidence for the existence of gluons. In some electron-positron collisions, three jets are produced. This was first seen at the PETRA collider in the DESY research center, Hamburg. The presence of these "three jet" events is expected in QCD, and is due to gluons. What happens is that very soon after they are produced (and while they are still feeling free!), either the quark or the antiquark radiates a high-energy gluon, Like the quarks, the gluon will fly away, and eventually turn into a jet of hadrons. As with the quarks, the properties of these jets can be calculated in QCD, and the predictions agree with the data. This is strong evidence that gluons are also real, even though they too can never be isolated.

Evidence for point-like quarks inside hadrons—rather than being produced in collisions—comes from experiments that scatter electrons off protons at high energies. A beam of electrons can be fired at stationary protons, as was done in the first such experiments, at Stanford University in California, or can be collided with a beam of protons to get even higher resolution. In either case, the experiment is acting as a super-high-resolution microscope, probing the insides of the proton—and, unfortunately, smashing it to pieces in the process. A huge amount of energy and momentum is transferred between the electron and the proton, usually via an exchanged photon. As usual, the energy and momentum set the wavelength of the photon, and thus the resolution of the microscope. When the energy is high, the wavelength is small, so the resolution is good.

At low values of energy, well below the crucial Lambda QCD scale, the microscope sees the whole proton. This is what happens if we perform the experiment back on Hadron Island. As the energy increases, the wavelength of the photon shrinks, and a smaller and smaller part of the proton is seen. Because of this, the probability of actually hitting the proton with the electron drops rather quickly.

However, in the first experiment in Stanford, and at other experiments since, it was seen that at some point, at energy transfers significantly above Lambda QCD, the scattering probability stops falling so fast. In fact, once you take into account the fact that the photon keeps on getting smaller as you go higher in energy, the scattering probability is almost constant. This is what happens if we perform the experiment now, on the Isle of Quarks.

This is what we would expect if there were tiny, point-like quarks inside the proton. The quarks are already infinitesimally small, so even when we shrink the wavelength, the part of them we hit can't get any smaller. The situation is very similar to Rutherford and his alpha particles, in discovering the nucleus way back west in Atom Land. The surprisingly high number of electron-proton scatters at high-energy transfers is the equivalent of the alpha particles bouncing right back from the gold foil that so shocked Rutherford and his team.

From observations like this, and from where they sit in the eightfold way, we know that protons are made of two quarks that carry two thirds of the electric charge of the proton and one quark that carries one third, adding up to a total of one, of course. However, the internal structure of a proton is much more complex than just three quarks sitting next to each other. The quarks are bound together in a small space by the strong force, which means they are exchanging gluons frantically. At the shortest distances we can observe, the proton is a complicated object containing lots of point-like quarks and gluons splitting and radiating. In a way, it is amazing that it is so stable, but it is. If protons ever decay, we have never seen one do so, despite very sensitive searches.

18

Flavors and Generations

Although similarly covered in railway lines, the Isle of Quarks is nowhere near as densely populated as Hadron Island. In contrast to the myriad hadrons, there are just six major cities dominating the landscape here. These are Up and Down, Charm and Strange, and Top and Bottom. Top is the heaviest, on the eastern tip of the island, and Down and Up are the lightest, farthest west.

The different types of quark are called "flavors."[35] We know from measurement of the hadrons they form, and what they decay into, that down, strange, and

35. The same word is used to describe the different types of lepton—electron, muon, and tau.

bottom all carry a charge of negative one third of the proton's charge, and the others carry positive two thirds. Strange was so called because it was first noticed as a constituent of some strange new particles seen in cosmic rays.[36] Charm solved some problems caused by the lonely strange quark, and bottom and top[37] mirror the names of down and up.

A remarkable thing about the Isle of Quarks is the distribution of the cities from west to east. Even though the quarks are, as far as we know, all infinitesimally small, their masses span a huge range.

Down and up are the constituents of protons and neutrons, and so between them make up the nucleus of every atom. They each have a mass of a couple of million electron volts, only four times heavier than the electron. However, that mass is swamped by the extra mass they get from the binding energy of the strong force, up to a few hundred million electron volts. (This is one reason the mass a quark would have if we could isolate it is not very accurately known.) In principle, these two quarks, along with the electron, could make up every atom in the universe, so it is something of a surprise that nature has copied them, at higher mass, farther east. There are three "generations" of matter—

36. Also in Manchester, England, in 1947—ten years after the first curry house, according to a report in the *Manchester Evening News*.

37. They have alternative names—beauty and truth—but they are too pretentious for most physicists.

three sets of particles following a similar pattern. Up, down, and the electron, along with a neutrino, constitute just the first of these.

The second generation consists of the strange and charm quarks, along with the muon and its neutrino. The second-generation quarks are higher-mass copies of the first, just as the muon is a higher-mass copy of the electron. The bare mass of the strange quark is about 5 million electron volts, and the charm is much more massive, at about 1.3 billion electron volts.

The third generation of quarks is heavier still, with bottom having a mass of 4.2 billion electron volts. The top quark is a ridiculously massive 172 billion electron volts, the heaviest of all the fundamental particles in the Standard Model. Because of this huge mass, the top quark decays so quickly that it never has time to be bound up inside a hadron.

The rather odd distribution of quark masses—with up, down, and strange fairly close, charm and bottom also not so far from each other in mass despite being different generations, and top being far away to the east—is not understood. The top quark is a real monster, nearly 200 times the mass of the proton, and yet unlike the proton it is supposed to be fundamental, an infinitely small point particle.[38]

38. This peculiar hierarchy of masses may well be one of the biggest clues to what lies beyond the Standard Model—certainly, the top quark plays quite a special role in many speculative new theories.

There are some tantalizing reasons to think that three generations is a special number, which may again be a natural consequence of a bigger, better theory beyond the Standard Model. A particularly interesting reason, for example, is that three generations is just enough to break the symmetry between matter and antimatter in the theory—between the northern and southern hemispheres on the map.

Touring the Isle of Quarks by train was a rapid and quite luxurious journey. Because the gluon carries color charge, there is even a rail bridge to Bosonia, a land that is firmly on our list of places to explore. Also in Bosonia, however, is a major airport. The planes we have occasionally seen overhead originate there, and we are offered tantalizingly rapid transportation to almost anywhere on our map. This is an opportunity we ought to take.

EXPEDITION V

The Isles by Air

Particles in pairs and frequent fliers – The locations of airports turns out to be important – Discussions with space aliens, physics in a mirror and the meaning of "south"

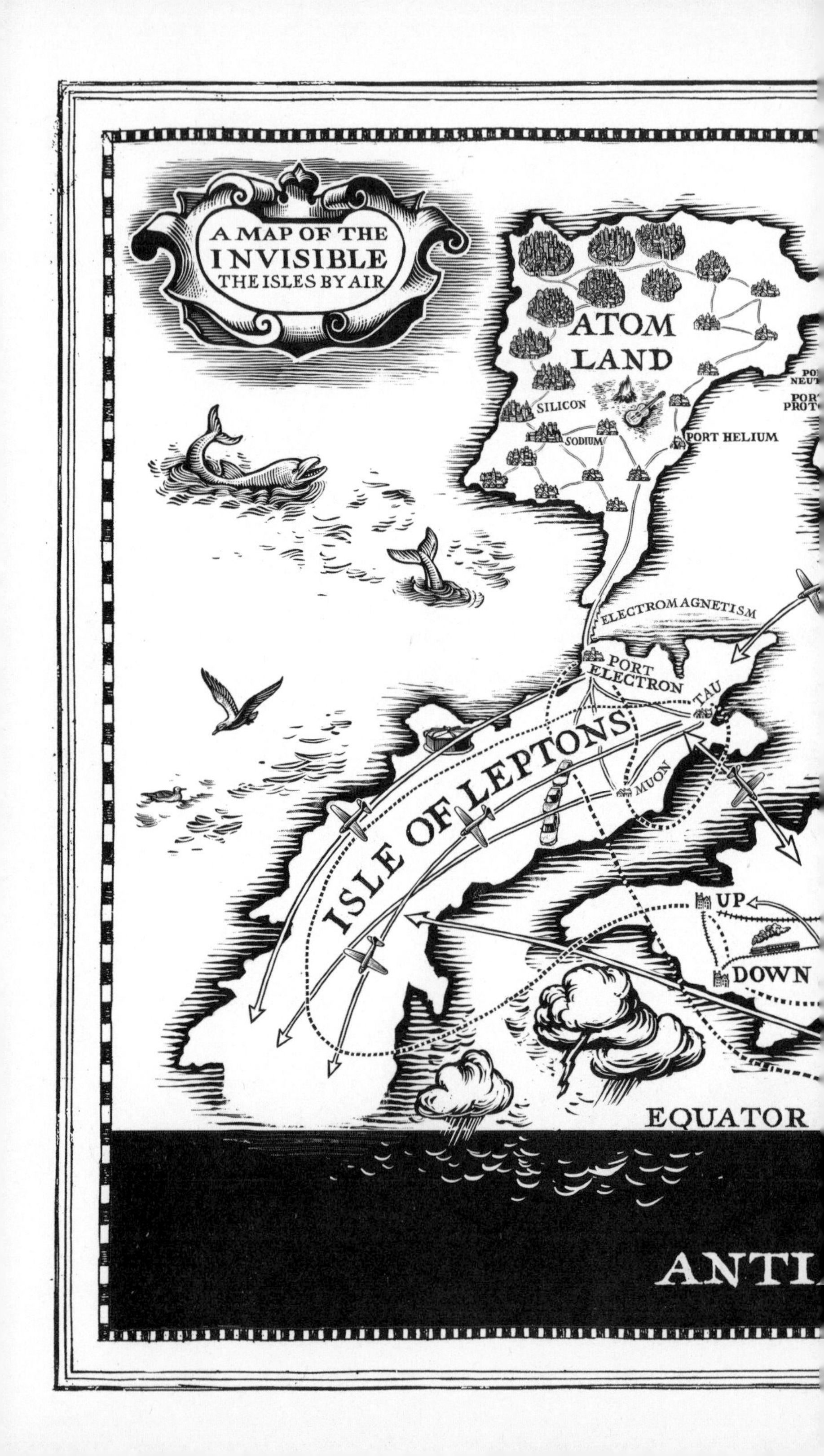
A MAP OF THE
INVISIBLE
THE ISLES BY AIR
ATOM
LAND
SILICON
SODIUM
PORT HELIUM
ELECTROMAGNETISM
PORT
ELECTRON
TAU
ISLE OF LEPTONS
MUON
UP
DOWN
EQUATOR

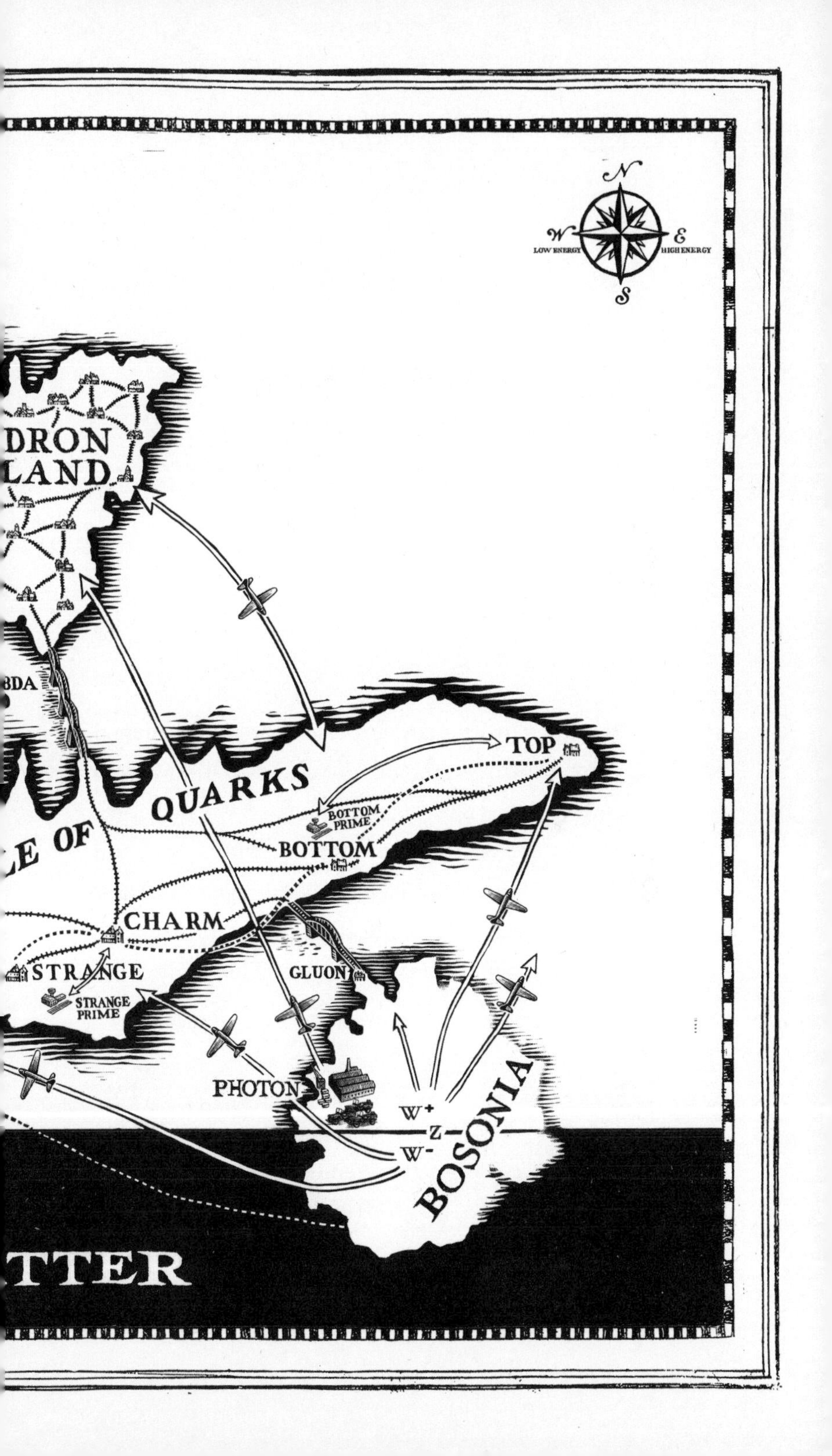
N
W
E
S
LOW ENERGY
HIGH ENERGY
DRON
LAND
BDA
QUARKS
LE OF
TOP
BOTTOM PRIME
BOTTOM
CHARM
STRANGE
STRANGE PRIME
GLUON
PHOTON
W+
Z
W-
BOSONIA
TTER

19

The Weak Force

The Standard Model incorporates three fundamental forces—three methods of transportation across the land we have been exploring. We have made use of two of them so far. Electromagnetism is our road network, connecting all electrically charged particles. The strong force is the dense railway system spanning Hadron Island and the Isle of Quarks. The airlines that we have just come across, at their hub in Bosonia, are the weak force.

The weak force does not manifest itself very obviously in everyday life. It is relatively easy to see the effects of electromagnetism—indeed we see *with* the photon, the boson that carries the electromagnetic force. And while the strong force may not be so visible, it plays an easy-to-appreciate, vital role in holding together the atomic nucleus.

It is a bit harder to point to the essential role of the weak force. It operates over such a short range that its influence is not felt directly, even at distances comparable to the radius of a proton. Because of the short distances involved, gaining an understanding of the weak force has required a journey farther east than most have undertaken so far—the airline hub is almost at the boundary of the map. Nevertheless, it is a crucial part of the Standard Model. Without the weak force, quarks cannot change flavor. That means that neutrons cannot transform into protons. That means that the Sun could not shine, since it relies on the fusion of four protons into helium, in which two of them become neutrons. So, weak though its influence is, this force is an essential component of our world.

The weak force does not really bind anything together in the way roads or railways do. It has no domain equivalent to Atom Land or Hadron Island. As the airline network on our map, it makes tenuous but ubiquitous connections to anywhere an airplane can land. The bosons that run the airline, carrying the force, are the W and the Z. The W has an electric charge, which can be positive (W^+) or negative (W^-). The Z is electrically neutral. Like gluons (which carry color charge) but unlike the photon (which is electrically neutral), the W and Z do carry the weak-force equivalent of charge, or "weak charge."

Every settlement on the map—every fundamental particle—has an airfield. All of them experience the weak force. But some are busier than others. And in particular, the airways between some *pairs* of particles are especially busy. This is one of several striking features of the weak force that sets it apart from electromagnetism and the strong force.

Electromagnetism—QED—tells us how an electron (or any particle with electric charge) can emit a photon. The theory of the weak interaction tells us how an electron (or any particle with a weak charge) can emit a Z or a W. It tells us that emitting a Z is much like emitting a photon. But when a W is emitted, things are somewhat different, because the W has an electric charge. So when an electron emits a W, it loses its electric charge, and turns into a neutrino. Port Electron and its mysterious neutrino have this dedicated air route between them; they form a pair, or doublet. Similarly, a muon can emit a W and turn into a muon neutrino, and a tau can do the same and turn into a tau neutrino—all by emitting a W boson. Note that electric charge is conserved—in each case the lepton's negative charge is carried off by the W, and the neutrino, in a rather inaccessible mountainous region of the Isle of Leptons, is neutral.

From the air, the same pattern is seen on the Isle of Quarks. Emission of a negatively charged W will turn a down quark into an up quark, because they are in a doublet together. It will also turn a strange quark into a charm quark, and a bottom quark into a top.

This ability to transform particles within doublets into one another is unique to the weak force, and it is important. Without it, for example, the top quarks created at the Big Bang would still be with us. There would be no way of getting rid of them.[39] The weak force is also responsible for the fact that isolated neutrons decay after a few minutes—in this case, the underlying process involves a down quark transforming into an up quark by emitting a W. The W then decays to an electron and an electron antineutrino.

39. Unless and until they meet with a top antiquark and are annihilated.

20

Parity, Helicity, and Chirality

The process by which a neutron converts into a proton is called "beta decay." It was in beta decay that the weak force sprang its biggest surprise.

The weak force interacts with all the quarks and leptons, but at the same time it only interacts with half of them, in a sense. Put another way, every city on the Isle of Leptons and the Isle of Quarks has an airport, but only half the inhabitants ever fly. Perhaps they are frightened or worried about climate change. Or maybe airport security is very picky. Whatever the reason, it is a puzzling fact. To work out what it means, we need to understand a concept known as chirality.

A chiral object is one that is not identical to its mirror image. It is also sometimes called "handedness." Some molecules are chiral, having a right-handed version and a left-handed version. This is true of many biological molecules; DNA is chiral, for example: The twist in the double helix goes in a particular direction. So are sugars. Usually only one of the two possible chiralities is actually usable by living organisms.

The particles of our map, too, are chiral, and only one of those chiralities is affected by the weak force.

Spinning particles can be divided into two classes based on whether their spin is pointing in the direction the particle is traveling, or opposite to it. This is called the helicity.[40] These are the only two possibilities for the helicity of a quantum particle with a spin of half a unit, such as the quarks and leptons of the Standard Model. If you imagine the particle coming toward you, you can think of the two helicities as corresponding to seeing the spin turning clockwise or counterclockwise, just as the twist of the DNA helix goes in one direction or the other. For a massless particle, these two helicities define the two possible chiralities of the particle.

This is strange behavior. The fact that there are two chiralities is not so surprising. Since we know that most particles have spin, there will be a handedness

40. From the same root as "helix." The tip of a spinning arrow will describe a helix as the particle moves along through space.

associated with the sense in which they spin, clockwise or counterclockwise, similar to the handedness of the twist in the double helix of DNA. What is surprising is that the W and Z bosons, which carry the weak force, "see" only one of those chiralities. A massless particle with its spin pointing in the opposite direction to the particle's direction of motion will feel the weak force; a massless particle that has its spin pointing in the same direction as its motion will be invisible to the weak force. The inverse is true for antiparticles: Those with spin pointing in the same direction as their motion feel the force; those with it pointing against do not.

The fact that in nature, plants and animals produce, and make use of, only one chirality of some molecules is presumably due to random chance—some common ancestor happened to make and use one of them, and the success of that ancestor has meant that the original choice has been "locked in" ever since. But why fundamental particles should have such a strong bias toward one of the two chiralities is far from obvious and is not understood within the Standard Model.

Whatever the reason for it, this chirality bias has profound consequences.

Before this property of the weak force was established, physicists had thought that the universe was completely symmetric under reflections in a mirror. Or to put it another way, we thought that the fundamental forces of physics did not distinguish between left and right. But the weak force does. If you imagine reflecting a spinning, massless particle in a mirror—an operation called a "parity inversion"—you flip its chirality, which will turn the weak force on or off. That can completely change the energy of the particle, or of any other particles nearby, and completely break the symmetry. It means "left" and "right" are no longer arbitrary labels—a difference between them is built into the structure of physics.

A good way of thinking about it is to imagine communicating remotely with a different civilization on a far planet. There would be many challenges, but let's imagine that eventually we establish a rudimentary common vocabulary, so we can chat reasonably fluently. We find they are also a carbon-based life form. They don't call it carbon, of course, but we agree on what atoms are, and that carbon is the one with six protons and six neutrons in the nucleus, with six electrons bound to it.[41] So we want to send them a box of chocolates as a gesture of goodwill. But we are worried that they may use sugar molecules with the opposite chirality to ours. If this were the case, they would not be able to digest our chocolates, which could be awkward, diplomatically speaking. How could we tell in advance?

41. Their map of physics should be the same as ours, after all, even if they use different words to describe it.

We could ask them, but we need to agree on a definition of what we call right or left, clockwise or counterclockwise. Images are no good because we can't tell whether we have decoded them with the correct chirality. There is no physical reference possible based on electromagnetism, because electromagnetism treats left and right identically. The same goes for the strong interaction. The weak interaction provides the only way.[42] What we might end up doing, laboriously, in our painstakingly built up common language, is exchanging instructions on how to build an atomic parity experiment.

In 1957, Chien-Shiung Wu and her team at Columbia University conducted the experiment that proved that the weak force does indeed violate parity symmetry. This seminal experiment involved taking atomic nuclei that undergo beta decay—a process governed by the weak interaction—and aligning them with a magnetic field, so that their spins pointed along the direction of the magnetic field. A reflection in a mirror does not affect the magnetic field—electromagnetism conserves parity—but it does flip the spin.

Measuring the electrons from the beta decay shows that they are mostly emitted in the opposite direction to that in which the spin of the nucleus is pointing. So

42. Apart from, presumably, saying, "See that funny-looking galaxy over there? That's on our left."

they point opposite to the magnetic field. This means we can agree with our space alien friend that "counterclockwise" corresponds to the direction of rotation of the spin of the electron as seen when looking along the magnetic field, and clockwise is the opposite, of course. From that, left and right can be unambiguously defined, and we can work out whether our chocolates will go down well or not.

The weak interaction changed our assumptions about physics forever. It connects every city on our map by air, even the inaccessible neutrinos. But not everyone in those cities uses the airport. All the right-handed bits are unable to fly. And despite being somewhat obscure in everyday life, the weak force has a surprisingly pervasive—indeed, critical—influence on why the universe is the way it is. It allows the Sun to burn, and the heavy generations of particles to decay into everyday matter. It also has other unique qualities, which we will investigate further as we hop around our map by air.

21

Mixed Messages

Studying the flight paths of the weak force across the Isles of Leptons and Quarks has revealed the fact that as far as the weak force is concerned, particles come in pairs, or doublets. Each lepton shares a doublet with a neutrino, the up quark is in a doublet with the down quark, and so on for charm and strange, top and bottom. The W boson provides direct connections between the two particles in a doublet, and this allows the top quark to decay to the bottom quark, and the neutron to decay into a proton by the conversion of a down quark into an up quark. So far so good. But as we become seasoned frequent fliers, more subtle effects start to become evident.

The top quark decays to a bottom quark because it is in the same doublet. But the bottom quark also decays to even lighter quarks, even though they are in different doublets. So do *all* the heavier quarks and leptons. The stable matter of the universe consists of up and down quarks inside every atomic nucleus, and electrons. This makes sense, because heavy particles will decay to lighter ones if they can, as we have noted before. But the key is "if they can." What is providing the routes westward that allow this? It is the weak force, again, but in quite a subtle fashion.

As we travel around, it becomes clear that the airports at which the weak force lands are not exactly in the cities. This isn't unusual; it's a fairly short journey from the airport near the strange quark to the city of Strange itself, and that's the way most of the traffic goes. However, it is possible to go directly to the down quark from there, too, or even to the bottom quark. It may be a bit of a long transit, and most passengers don't do this; they fly to the Strange airport and go straight into town. But it is possible to go instead to Down or Bottom directly from Strange airport, and some do so.

In terms of our map, there are two ways of looking at the settlements on the Isles of Leptons and Quarks. From the point of view of the inhabitants, the most important thing is the cities. On the Isle of Quarks, these would be Up and Down, Charm and Strange, Top and Bottom. Each of these quarks has a definite mass, with top being the heaviest and down and up the lightest.

But from the point of view of an airline pilot, or a W boson of the weak force, a more sensible way to think about it might be in terms of the airports. And for Down, Strange, and Bottom, the airport is not in quite the same place as the city. They are nearby, but as we've noted, one can fly into the airport near Down and travel directly on to Strange or Bottom. To make it clear they are near towns, the airports are called Down-prime, Strange-prime, and Bottom-prime. The airline connects these locations, rather than the cities of Down, Strange, and Bottom themselves.

This is crucial to understanding why the stable matter of the universe only contains up and down quarks. To get at the reality behind this airport placement, we need to return to the idea of a quantum state. A quantum state is the object that carries all the information about a particle, or set of particles. It is possible to extract that information in several different ways that sometimes seem contradictory. Different ways of observing a quantum state can even give different answers as to what kinds of particles it describes. In terms of the map, looking at the island from the point of view of the cities or the airports is the equivalent;

you still cover the whole island. The Standard Model really does the same thing with quarks. If you have a quantum state describing some quarks, you can analyze it as being a certain amount of down, a fraction of strange, and a bit of bottom. This makes sense, and since those quarks have different masses, that is how they will appear in hadrons, too. However, that is not how the weak force (awkward as ever) sees them. The weak force splits the quantum state a different way, into a different set of "primary" particles. These are the different flavors of quark that can be produced when a W boson decays, and we call them d′ (down-primed), s′, and b′. They don't have definite masses, but they are the particles the weak force sees. We don't know why it does this (any more than we know why it only sees left-handed particles!), but that's the way it is, and it is a good thing, too, as it has important consequences.

The strange-prime quark is nearly the same as the strange quark, but it has a small admixture of down and bottom, too. It is as though we take a selection of people in the city of Strange and ask them where they live. They all say Strange. But then we ask them what airport they landed at. Sensibly enough, most of them say they flew into the nearby Strange-prime. But a few came into Bottom-prime, and a few came into Down-prime. Likewise, if we survey a plane full of passengers landing at Strange-prime and ask them which city they are heading to, most will say Strange City, but a few will say Bottom and a few will say Down. When a quark travels through space and forms a hadron, a particular mass is selected; this is the city it belongs to. But a W boson that decays will not produce pure strange quarks—it will produce strange-prime quarks, which is the airport it lands at. And the strange-prime quark has a small probability of being a down quark, or a bottom quark—that is, some of the people landing at Strange-prime airport actually live in the cities of Down or Bottom. This is the reality behind what we observed on our map—that even if we land at the Strange-prime airport, we could (sometimes, not often) travel straight to the city of Down or Bottom, instead of Strange.

These crucial small probabilities are what allow the weak force to communicate between the generations. This is what allows the heavier particles of the second and third generation to eventually (or sometimes quite rapidly) decay to the lightest, first generation. This is how the weak force provides a route west from Top, Bottom, Charm, and Strange all the way to Up and Down. And that is why we don't have a universe filled with atoms containing charm or strange quarks inside their nuclei.

22

North from South

On our map, the northern hemisphere is matter, and the southern is antimatter. We have been traveling almost entirely around the north, but the south would look very similar, if not identical. Or maybe we are traveling in the south, really, not the north? Could we even tell? Does the difference matter, or is it just a convention? Now that we have an airline at our disposal, we can take the view from 30,000 feet.

In a sense, very often, it does not matter at all. According to the Standard Model, the electromagnetic force between an electron and a proton is exactly the same as the force between their antiparticles, a positron and an antiproton. So while we would obviously notice (briefly, before we were annihilated in a burst of photons) if half the world were made of matter and the other half of antimatter, would we be able to tell if, in an instant, all matter were swapped for antimatter, and vice versa? From measurements of electromagnetism, or the strong force, or gravity, we could not—at least, as far as we know.

This is similar to the conundrum we had with parity, and trying to explain left and right to our space alien friend. We would have very similar problems trying to work out whether they were made of matter or antimatter. This is an important question to resolve before setting up a diplomatic mission. Since matter and antimatter annihilate, an error would be catastrophic for any diplomatic visit.

The operation of swapping all particles for their antiparticles and vice versa is called "charge conjugation." We saw that the weak force violates a similar symmetry, the left-right symmetry of parity, since it only affects left-handed particles and right-handed antiparticles. This means that it also breaks charge-conjugation symmetry, since if we charge-conjugated a left-handed electron, we would get a left-handed anti-electron. The former experiences the weak force, but the latter does not. In terms of our map, in the southern hemisphere a different half of the population is afraid to fly.

However, if we simultaneously charge-conjugate everything and at the same time reflect it in a mirror, we may be OK. The left-handed electron (which feels the weak force) becomes a left-handed positron (which does not) under charge-conjugation, but then becomes a right-handed positron under reflection (which again feels the weak force).

Thinking back to the parity discussion, and trying to communicate with a distant space alien to define left and right, we have a problem. If the space alien is made of antimatter, not matter, we will get the wrong answer. Our neat trick with the Wu experiment doesn't work, because if they are made of antimatter, their magnetic field will point in the opposite direction, so there are two configurations that give the same result. Either they are made of matter and we use the same definition of handedness, or they are made of antimatter and use the opposite handedness definition. We can't tell these two scenarios apart using the Wu experiment.

The weak force again provides a way out of the impasse. There is still a measurement that we can perform to decide whether or not it is safe to meet the alien. The weak force violates the combined symmetry of charge-conjugation and parity in a very subtle fashion, a way that is intimately related to that fact that we have three generations of matter, and to the way the weak force connects the primed objects (the airports) rather than the massive quarks themselves (the cities).

We have to quantify the relationship between the airports and the cities—between the primed and unprimed versions of the quarks. We can do this using a matrix, an array of numbers that encodes how much down, strange, and bottom are mixed into each of the down-prime, strange-prime, and bottom-prime states. This matrix will tell you exactly how likely a passenger landing at the Down-primed airport is to go to each of the Down, Strange, or Bottom cities—there is a number in the matrix for each possibility.

The presence of this mixing matrix is what introduces the possibility of the violation of the combined charge-conjugation and parity symmetry. That is what we need to allow us to communicate with our space alien friend and be sure we agree on what is matter and what is antimatter *and* what is left and right. The possibility arises as follows.

The matrix can contain a term that affects the phase of the quarks' quantum state. Remember, from the early preparation for our journey, that the phase of a wave is just the stage it is at in its oscillation, the position of Feynman's little rotating arrow. Phase itself is not physically measurable, but differences in phases are. This is why violation of the charge-conjugation and parity combined symmetry is a subtle effect.

The matrix allows mixing between different flavors of quarks, and when they are bound together in hadrons, this may also allow mixing between particles and antiparticles.

For example, there is a hadron (called a "neutral kaon," one of the many hadrons we whizzed past on Hadron Island) that contains a down quark and a strange antiquark. Since the down quark has a charge of negative one third and the strange antiquark has one of positive one third, the hadron has a total charge of zero. The antiparticle of the hadron also has zero charge but is made up of a down antiquark and a strange quark. By exchanging W bosons between the quarks, which involves the mixing matrix, the neutral kaon can transform into its antiparticle, and back again, "silently," oscillating between particle and antiparticle as it travels through space.

So far, so weird, but this in itself is not a violation of the combined symmetry.

In fact there are two possible ways for the kaon to transform between particle and antiparticle—two different ways the W bosons can be exchanged. So the physical result involves the sum of both these ways. And the phase term in the matrix affects these two ways differently. This introduces a phase *difference* between the two, which can be measured by precise detection of the particles that are produced when kaons decay.

It has been measured this way, and was found to be different when the hadron was going from particle to antiparticle than when it was going from antiparticle to particle.

That's where the violation of the combined symmetry comes in. Not only is it possible in the Standard Model, it actually happens. It is a small but very significant effect.

It would take a long conversation to explain to a space alien what special kinds of hadrons are needed, how to make them, and what properties to measure. But in the end we could do it, and we could then be sure if they were made of matter or antimatter, and then also agree on left and right.

We have discovered, by looking carefully enough at the airline traffic of the Isle of Quarks, that nature does not respect the symmetry between matter and antimatter, or between left and right, or the combination of the two. There is a real difference between the physics of the northern and southern hemispheres on our map.

There are several odd things about this. Firstly, only the weak force violates these symmetries. Why would that be? Once you have established that there can be differences, it would be very easy, for example, for the strong force to show them, too, but it does not appear to. Once you know that one force of nature violates the symmetry, the fact that the strong force conserves it is actually a puzzle. It is one of the possible hints as to what may be going on in the far east of our map.

Secondly, the combined symmetry can only be violated in the matrix if there are three or more generations of matter. Two would not be enough. So nature seems to have the minimum number of generations required to allow this distinction between left and right, and matter and antimatter, to be drawn. Those extra quarks and leptons may be there for a reason after all! Yet within the Standard Model, this looks like a coincidence. Maybe it is another clue to a better theory, again in the far east, where this is all obvious and necessary?

Finally, it is obvious that the universe around us is not symmetric between matter and antimatter, just as it is obvious on the real globe that the southern hemisphere is different from the north. There is much more matter than antimatter in the universe. Yet the symmetry violation in the Standard Model is such a tiny effect that no one knows how to build a theory in which it leads to such a big imbalance in the universe today. This again may be a clue to wild and wonderful things off to the east.

EXPEDITION VI

The Remote Neutrino Sector

A short but arduous trip around some rough terrain – A budget airline – More mixing – When is a particle an antiparticle, too?

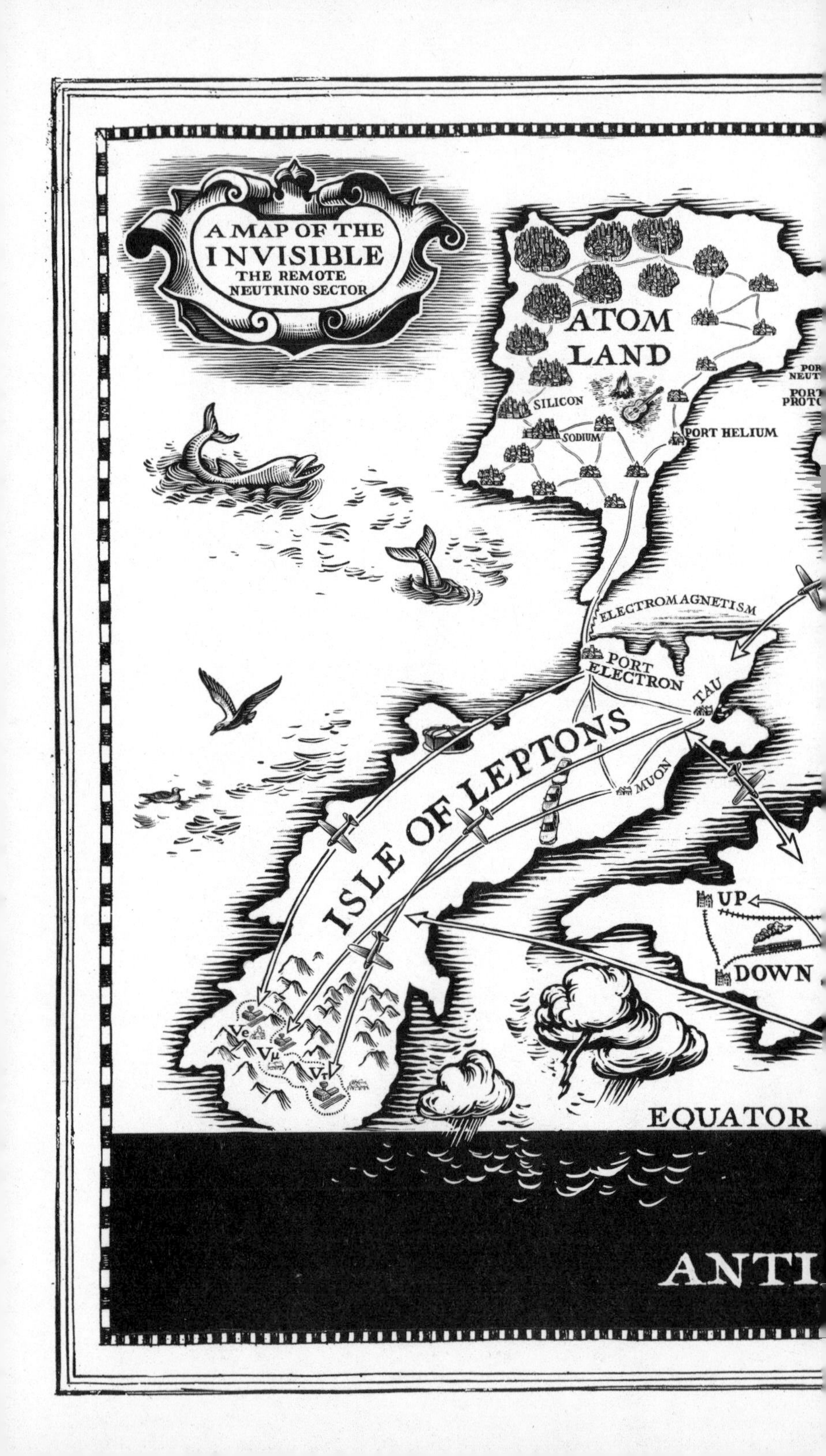
A MAP OF THE
INVISIBLE
THE REMOTE
NEUTRINO SECTOR
ATOM
LAND
SILICON
SODIUM
PORT HELIUM
ELECTROMAGNETISM
PORT
ELECTRON
TAU
ISLE OF LEPTONS
MUON
UP
DOWN
EQUATOR
ANTI

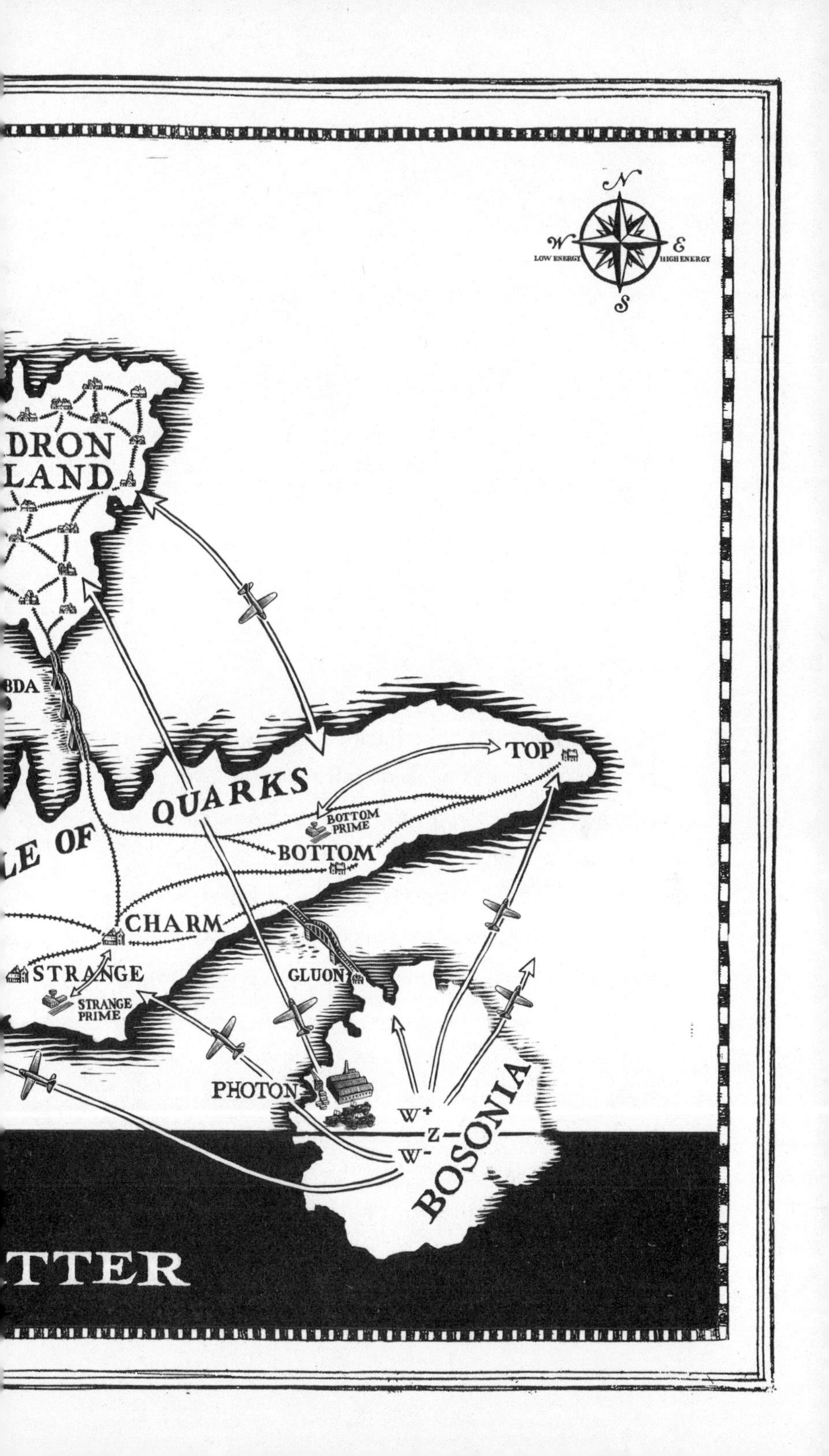

N
W
E
S
LOW ENERGY
HIGH ENERGY
DRON
LAND
BDA
QUARKS
LE OF
TOP
BOTTOM PRIME
BOTTOM
CHARM
STRANGE
STRANGE PRIME
GLUON
PHOTON
W+
Z
W-
BOSONIA
TTER

23

Massless Matter?

Port Electron, on the Isle of Leptons, was our very first stop, and subsequently we discovered and visited the muon and the tau lepton, important features on the same island. These places are easy to get to by sea or road—they are electrically charged, so electromagnetism gives us easy access. But in studying their airports, learning about the weak force, we have several times come across rumors of neutrinos. We have seen departure boards listing flights to Electron Neutrino, Muon Neutrino, and Tau Neutrino, and we have seen people getting on and off these flights, so presumably the destinations exist. But so far we know next to nothing about these mysterious settlements. They are inaccessible by road or rail—the electromagnetic and strong forces just don't go there—and while the weak force can get us there by air, so far we have had little more than a high-altitude flyby. It was enough to see that they are a crucial part of our landscape, and we have seen that they are important players in beta decay and the reactions that keep the Sun burning, but little else. The difficulty of the terrain notwithstanding, it is time to rectify that.

Neutrinos were first postulated in 1930, by Austrian[43] theoretical physicist Wolfgang Pauli.[44] Famously and rather shamefacedly, he said, "I have done a terrible thing, I have postulated a particle that cannot be detected."

The reason he did this had to do with beta decay. As we have seen, in beta decay, a decaying nucleus emits an electron. These electrons can be detected and their energy can be measured. One of the basic laws of physics is that energy and momentum are both conserved. The totals before the decay will be the same as the totals afterward. We can use that to make some predictions.

Imagine the nucleus is stationary before it decays. The total momentum is zero. So after the decay, the total momentum must also be zero, since total

43. And Swiss. And American.
44. He also developed the exclusion principle we discovered in Atom Land.

momentum is conserved. This means that if the electron moves off in one direction, the nucleus must recoil in the opposite direction to cancel out the electron's momentum, leaving the total as zero.

Likewise, energy is conserved. The electron and the recoiling nucleus have some kinetic energy, and the final mass of the nucleus will have reduced slightly during the decay, by an amount exactly sufficient to provide the mass of the produced electron and that kinetic energy.

If you put all that together, it is possible to solve the conservation equations and predict exactly the unique momentum the electrons must all have. Other types of radiation (alpha and gamma) bear this out. Alpha particles and gamma rays have a fixed energy for a given nuclear decay. But for beta radiation, the answer is . . . wrong. The electrons have a spread of momenta and energies, always less than the prediction. This is a problem.

There are only two options. Either energy and momentum conservation don't work in beta decays (which was proposed as a solution by Danish physicist Niels Bohr), or we are missing something in the decay. Pauli went for the latter, inventing the neutrino, which could carry off variable amounts of energy and momentum and thus balance the nucleus and the electron. Energy and momentum would still be conserved, but the electron could now have a range of energies, as observed, with the maximum occurring when the neutrino had almost zero energy, and the minimum occurring when the neutrino carried away the maximum it possibly could. All the energies in the range would be consistent with conservation of momentum, and they would be distributed randomly according to the probabilistic nature of quantum mechanics.

From our point of view as explorers, we are sitting in the arrivals hall of the airport near the city of Up. We see more passengers landing than are emerging in the arrivals hall. Either some passengers are vanishing (which would be very disturbing), or they are transferring to connecting flights onward to the neutrino.

Thankfully, the latter is the case. Pauli was right about the existence of the neutrino. But he was wrong about the impossibility of detection. Getting to the Neutrino Sector of the Isle of Leptons is very difficult, but it can be done.

In the original conception of the Standard Model, neutrinos were in a unique position—the only massless matter particle. The reason for this had to do with the weak force, the only force they experience.

As we saw studying the airlines on our map, the weak force interacts with only one chirality: left-handed particles and right-handed antiparticles. Since the only force the neutrinos experience is the weak force, this means that the right-handed neutrino and the left-handed antineutrino are *totally inaccessible* in the Standard Model. There is no access by road, rail, or air. Electromagnetism, the strong force, and the weak force do not connect to them! It would be less disturbing if such particles didn't even exist—after all, how could we even tell if they did?

How does this relate to the masslessness of neutrinos in the original Standard Model? There is a fascinating connection involving both relativity and quantum field theory that we will have to negotiate in order to access and understand the Neutrino Sector properly.

When looking at the strange asymmetric behavior of the weak force, and the way it affects only left-handed particles and right-handed antiparticles, we only discussed massless particles, for good reason. Because once a particle has mass, the definition of left- and right-handed particles gets a little more complicated. The helicity—the direction of spin relative to the direction of motion—no longer defines the chirality. It is still possible to define chirality as the feature that the weak force cares about, but for a massive particle, chirality is no longer exactly the same thing as helicity.

If you think about it, this has to be true, because the helicity of a particle depends upon the relative speed of the observer and the particle. If we chase a positive-helicity particle and overtake it, the spin does not change direction, but the relative motion does, so the helicity would flip over. This is the same effect as if we look at the hands of a clock from behind its face—they go counterclockwise.

So by overtaking a particle, we alter the helicity. If this changed the weak force, that would be unambiguously observable, and it would give us a way of measuring absolute speed, in violation of relativity. We could use the strength of the weak force to define the absolute direction of travel of particles, without reference to anything else. Just as bad, say we catch up to the particle but don't overtake it, so that relative to us it is stationary. For a stationary particle, there is no direction of motion, so no defined helicity. So what does the weak force do?

What does happen is that when I catch up with and overtake a particle, the helicity changes, but the chirality does not change, and the weak force does not change.

Massless particles all travel at the speed of light, and it is not possible to bring them to rest or overtake them. In this case, helicity and chirality are identical. For massive particles, the helicity is still correlated to the chirality, but is not identical to it.

The upshot of this is that for massive particles, a particle with a definite helicity does not have a definite chirality, and conversely a particle with a definite chirality does not have a definite helicity. So if we create a pure beam of particles with a definite helicity, it must contain both chiralities. For a massive particle, both chiralities must exist in nature, but for a massless particle, we can get away with having only one chirality.

Going back to the idea of a massless neutrino then, if the neutrino is massless we can build a theory containing only the left-handed neutrinos and right-handed antineutrinos. Only the particles that actually interact with the Standard Model forces need to exist. The right-handed neutrinos and left-handed anti-neutrinos need never even exist.

Appealing to Occam's razor, *non sunt multiplicanda entia sine necessitate*,[45] or the simplest answer is usually the correct one. The Standard Model was originally constructed so that it did not contain right-handed neutrinos or left-handed antineutrinos. They were a blank space on the map. And this meant that the neutrino had to be massless. What we are saying is that in the so-far-uncharted neutrino badlands, which can only be reached by air, only the airports exist.

But is this really true?

45. Entities must not be multiplied beyond necessity.

24

The Standard Model Is Dead—Long Live the Standard Model!

At last we take a flight to one of the neutrinos—the Electron Neutrino—and start to explore the surrounding territory. According to the Standard Model, this tiny airport, and similar ones for the Muon Neutrino and Tau Neutrino, should be all there is to see. But doubts creep in with further explorations.

Apart from the general weirdness of a particle that can only exist with one "handedness," a major worry for physicists exploring this territory was the "solar neutrino problem," first observed in experiments led by American physicists Raymond Davis Jr. and John Bahcall in the late 1960s and early 1970s.

The essence of the problem with solar neutrinos as Davis and Bahcall understood it was that there weren't enough of them. Our knowledge of the Sun and nuclear physics had led to a prediction that a certain number of neutrinos should be being produced by the fusion reactions in the Sun. These would be "electron neutrinos," meaning they are produced in association with electrons, as in the radioactive beta decay of unstable nuclei.

Since they only experience the weak force, neutrinos interact very rarely, so they are very hard to detect. One thing they can do, however, is scatter off nuclei and produce electrons. This involves the exchange of a W boson, which carries electric charge, and is called a "charged-current" interaction. It is sort of the inverse of the process by which the neutrino is originally created.

Bahcall and Davis predicted how often this should happen—in the specific case of neutrinos from the Sun hitting chlorine nuclei and producing argon plus an electron—and designed and built an extraordinary experiment to measure it.

The experiment used 440 tons of perchloroethylene, a chemical[46] commonly used in dry-cleaning fluid. Neutrino reactions within this substance produce argon by turning one of the neutrons in a chlorine atom into

46. C_2Cl_4—two carbons with four chlorines.

a proton. These reactions are so rare that the tank containing the fluid had to be buried underground to shield it from other particles. Davis used the Homestake gold mine in South Dakota. The argon produced was an isotope with nineteen neutrons inside the nucleus along with its eighteen protons, which is unstable but decays quite slowly, with half of the atoms in any sample decaying in five weeks. Davis collected the argon every few weeks and counted the argon by detecting its decays. He found that the amount of argon was lower than predicted by Bahcall's calculations based on our knowledge of the Sun. The number of neutrino interactions looked to be lower than expected by a factor of three.

This meant that something was wrong. Either we didn't understand the Sun, or we didn't understand nuclear physics, or the experiments included a mistake, or the Standard Model of particle physics was wrong.

The results were so groundbreaking that many people suspected the problem must lie with the experiment. But Davis continued collecting data and refining his experiment, and different experiments over several years repeated the measurement. The discrepancy remained.

If the problem were with the Sun, we could be in very serious trouble. The fusion reactions that power the Sun produce neutrinos and photons. It is the photons that keep us warm—they are sunlight! The neutrinos almost all pass straight through Earth. A neutrino produced inside the Sun will reach us in about nine minutes, traveling at the speed of light. But a photon will be delayed in the Sun, bouncing around among the plasma, being absorbed and re-emitted many times over thousands of years before eventually reaching the surface. Once there, it will also get here in about nine minutes. One possible reason for the lack of neutrinos was that there were fewer fusion interactions going on in the heart of the Sun. Maybe, scientists thought, the experiments indicated that the Sun was running out of fuel, a lack that has not affected the light and heat we see from it yet because of the time lag in photons getting to the surface. But the Sun's days, or years at least, would be numbered.

If the problem were indeed with particle physics, the solution would be less threatening, but intriguing nonetheless. This solution is suggested by the fact that the experimentally measured number of neutrinos was too low by a factor of three.

We know there are three different kinds of neutrinos in the Standard Model: the electron neutrino, the muon neutrino, and the tau neutrino, each one in a doublet with its partner charged lepton, and each interacting with that lepton via the W boson of the weak force.

These are the airports, the only way into the Neutrino Sector of the Isle of Leptons.

However, the experiments were able to detect only the first kind, the electron neutrino. The neutrinos from the Sun should be produced as electron neutrinos, because that is how nuclear physics told us those fusion reactions work. But what if they changed on the way from the Sun to Earth? What if the neutrino types mixed themselves up, so that by the time it reached us the beam of neutrinos contained an equal amount of all three types? That would neatly explain the discrepancy, since two thirds of them would be invisible to the detector.

There is already a way of changing the type, or flavor, of matter particles in the Standard Model. We saw it on the Isle of Quarks. The definite mass versions of quarks do not exactly line up with the versions that are produced by the weak force—on our map, the airports are not in the same place as the cities—so quarks can change flavor as they travel, and oscillate between flavors. Could the same thing be happening with neutrinos? If the neutrinos were to mix flavors on their way from the Sun to Earth, then on average there are likely to be equal amounts of the three kinds arriving at the detectors. Two out of three of these would be completely invisible, and thus the detector would measure an answer that was too low by a factor of three, as observed.

Following the situation we observed on the Isle of Quarks, we know how to handle this. We need a situation where the airports don't exactly line up with the cities, and we need to quantify the effect using a mixing matrix for the neutrinos, similar to the one we have for the quarks. But in order to do this, we need three neutrino cities to exist, just as the quarks have cities distinct from the airports.

The distinguishing feature of the cities, though, is their definite and distinct masses. This won't work if all the neutrinos have the same mass! If we are to use neutrino mixing to solve the solar neutrino problem, the neutrinos have to have masses different from each other. And specifically, that means at least two of them (and probably all three) must have non-zero masses.

We know that has a serious impact, though. Following our examination of helicity and chirality, a non-zero neutrino mass means that the left-handed neutrino can also become right-handed, and we can no longer get away with having only left-handed neutrinos and right-handed antineutrinos. There has to be a right-handed neutrino. We must introduce a new particle that does not interact with any of the forces of the Standard Model!

Sorting out what is really going here requires data from a definitive experiment: the definitive expedition to really work out how the land lies. The Sudbury Neutrino Observatory—SNO—was just that.

The key idea was to design an experiment that could detect neutrinos of *any* flavor by measuring the "neutral current" interaction. In this interaction, a Z boson is exchanged instead of a W. Because the Z carries no electric charge, the neutrino breaks up a nucleus, but remains a neutrino. No electron (or muon or tau) is produced. The only sign that it was there would be a few hadrons, the shattered products of the broken nucleus.

Finding these hadrons required detectors of unprecedented sensitivity. To confidently detect the tiny signal of a broken nucleus, all other extraneous radioactivity has to be reduced to the lowest possible level, and measured precisely. To do this, the SNO collaboration borrowed 1,100 tons of "heavy water" from the Canadian atomic energy program. Heavy water is just like normal water, with two hydrogen atoms bound to an oxygen atom, but one or both of the hydrogen atoms are replaced by a heavy isotope of hydrogen—deuterium—which has a nucleus of one neutron and one proton, as opposed to a normal hydrogen, which has just a single proton. This makes the water 5 to 10 percent heavier than normal water. Other than that, it is just like normal water. You could swim in it or take a sip of it, although experiments with mice indicate that more than about 20 percent of heavy water in your drink is very bad for you and more than about 50 percent is probably fatal.

The SNO heavy water was stored in an acrylic vessel surrounded by normal water, and watched carefully with an array of highly sensitive photon detectors.

Three different things can happen in the detector at SNO when a neutrino interacts with heavy water. One thing is that the neutrinos convert a neutron into a proton—the charged-current interaction—and produce an electron. Only electron neutrinos can do this. Or they can break the deuterium nucleus up into a neutron and a proton—the neutral-current interaction, so-called because no electric charge is exchanged. This reaction is possible regardless of the neutrino flavor—electron, muon, and tau neutrinos all do it. Neutrinos may also scatter off an electron via the neutral-current interaction; again, all neutrino flavors can do this, although electron neutrinos are about six times more likely to do so than the other flavors.

The different kinds of collisions leave different patterns of light in the detectors. Counting the rates for the neutral-current interactions told the physicists at SNO the total number of neutrinos coming from the Sun, regardless of whether they oscillated between different flavors or not. The rate of charged-current interactions of electron neutrinos told them how many electron neutrinos were arriving at Earth. The difference between the two measurements told them how many neutrinos had changed flavor on the way.

The result was spectacular. The total number of neutrinos arriving at Earth from the Sun was bang on the expectation; the Sun was fine, and we had properly understood the fusion reactions driving it. The number of electron neutrinos was too small by a factor of three. Those early, immensely challenging experiments by Davis, et al., were vindicated, at very high precision. The Standard Model had to be changed—neutrinos are definitely changing flavor between the Sun and Earth. Therefore, they have mass.

In fact, while SNO settled the solar neutrino problem once and for all, another experiment, Super-Kamiokande in Japan, had already made big inroads into the neutrino wilderness of the Isle of Leptons, and shown in a different measurement that neutrinos had mass.

The Super-Kamiokande consists of an underground pool of more than 50,000 tons of purified water surrounded by photon detectors. The key measurement it made was of muon neutrinos. These are produced when high-energy particles hit the upper atmosphere of Earth. The particles come from all directions, and the neutrinos will pass through the atmosphere, and indeed pass through Earth. So in this quiet, underground pool under a mountain in Japan, muon neutrinos were expected to arrive from all directions equally. They did not. Super-Kamiokande detected only about half as many neutrinos arriving from below as from above. This suggested that, in the longer journey through Earth they had to make compared to neutrinos from directly above, some of the muon neutrinos were changing into tau neutrinos—to which Super-Kamiokande was insensitive. Subsequent detailed analysis of the angular distribution confirmed this. This kind of change in neutrino type can only happen if the neutrinos of definite mass (the cities) are distinct from those associated with a particular flavor of lepton (the airports), as we already saw on the Isle of Quarks. For this to be the case, neutrinos must in fact have a mass that is not zero.

This therefore showed that neutrinos did have mass; and SNO showed that this (in the form of electron neutrino flavor changing) was indeed the solution to the solar neutrino problem. Considering the difficulty of getting to the Neutrino Sector of the Isle of Leptons, this was amazing progress. It was the first, and so far only, fundamental change forced upon the Standard Model since its inception. It means that the neutrino territory is more complicated and interesting than we thought, and that there might be much more to be learned from exploring it.

25

Neutrino Badlands

So let us exit the tiny airport we landed at and head out to explore the surrounding Neutrino Sector on foot.

The fact that neutrinos have different masses, and mix, indicates that there is indeed a mixing matrix for leptons, just as there is for quarks. So we should expect that the airports on the Isle of Leptons are somewhat displaced from the population centers, just as the Down-prime, Strange-prime, and Bottom-prime airports are displaced. This is the case, but the picture is rather different.

On the Isle of Quarks, the two different versions of down quark, with and without the prime, are close enough that it is obvious that they belong together. However, for the neutrinos, the mixing matrix is very different. To explain the observations, the mixing must be much larger than on the Isle of Quarks.

The only way we have of fully exploring the Neutrino Sector is by air. The weak interaction connects three airports: Electron Neutrino, Muon Neutrino, and Tau Neutrino. SNO and Super-Kamiokande—and other experiments at accelerators and reactors that have followed them—have identified the fact that these airports are not the whole story; there are cities that the airports serve. These are versions of the neutrinos with definite mass. But the weak interaction seems to be running a budget airline into the Neutrino Sector. The cities are miles away from the airports, and in fact we still aren't even sure where they are exactly. This is not a very satisfactory state of affairs.

There is an interesting possibility here, though. Remember, the mixing matrix for the quarks breaks the combined symmetry of charge-conjugation and parity, introducing a real distinction between the northern and southern hemispheres of our map, matter and antimatter. The matrix we have just been forced to introduce for neutrinos also includes this possibility.

In the quark case, this effect has been measured. Neutrinos are harder to study, because they interact too rarely, and this aspect of the mixing matrix has not yet been measured. If it turns out that there is a large effect, it could help explain how the universe evolved to be mostly made of matter, not antimatter.

Our exploration of the Neutrino Sector of the Isle of Leptons has been very fruitful, but there might be still more out there. There are still other explorers braving the wilderness, trying to establish whether the neutrino is even the same kind of particle as all the other leptons, whether the Dirac equation even applies to it at all, or whether it dances to a different drum. The neutrino has a startling possibility open to it, one that is unavailable to the other matter particles. It might be its own antiparticle. We might have the Isle of Leptons wrong. The wilderness of the Neutrino Sector might extend right down to the equator—we still do not know, but we are trying to find out.

Matter is the same as antimatter except that all the "charges"—the quantum numbers that determine the interactions with the forces—are the opposite. So the electron is negatively charged and its antiparticle, the positron, is positively charged and therefore obviously different from the electron. Likewise, for a quark carrying a color charge (call it "blue"), the corresponding antiquark will carry the opposite charge ("antiblue," or "yellow" if you want to stick with the analogy to real colors).

Before we knew they had mass, we knew that neutrinos don't feel the electromagnetic or strong forces. And even the weak force acts only on the left-handed neutrinos. Back when neutrinos were thought to have zero mass, the Standard Model contained only those left-handed neutrinos. But now that we know they have mass, that doesn't work anymore. The mass implies that right-handed neutrinos must exist, too, mixed up with the left-handed ones. These right-handed neutrinos have zero charge with respect to all the forces—no electric charge, no color charge for the strong force, and no weak charge. So inverting the charge makes no difference—negative zero is the same as positive zero! Therefore there is a possibility—and many theorists would say a high probability—that the right-handed neutrino and the left-handed antineutrino are the *same particle*, with just the helicity flipped.

This kind of particle would appear in the equations of physics in a different way from the quarks and the other leptons; in particular, the way their mass appears is different.[47] The Neutrino Sector is currently being scoured for such a so-far-hypothetical particle. The search focuses on some very rare and special nuclear decays that, if observed, would have huge implications for physics and cosmology.

47. The relevant equations were first written down by Italian physicist Ettore Majorana, and so this kind of particle is often called a Majorana particle, leading to unexpected drug references in many physics presentations.

The rare decay in question is "neutrinoless double-beta decay." We have come across beta decay already. It is the process in which either a neutron becomes a proton or a proton becomes a neutron inside an atomic nucleus, changing the atomic number by +1 or −1 respectively. The process was first observed around the end of the nineteenth century, and the "beta particle" that is emitted is now known to be an electron in the first case, and a positron in the second.

Remember, beta decay provided the first indication of the existence of neutrinos. The emission of a neutrino, even if unobserved, allows the electron or positron that is emitted to have a spectrum of energies rather than a fixed value, as measured in experiments.

As the name suggests, in double-beta decay two neutrons or protons transform at once and two electrons or positrons are emitted, along with two neutrinos. This sounds unlikely, but for some nuclei the energy balance is such that this is the only way they can decay. The process is rare, but has been observed and measured in several different isotopes.

The existence of nuclei that undergo double-beta decay opens up a new and intriguing possibility. Emitting a particle is in many senses the same as absorbing an antiparticle. So if the neutrino can mix with its antiparticle, the same neutrino could be both absorbed and emitted in a double-beta decay, meaning that overall no neutrinos go in or out. This would be neutrinoless double-beta decay. In this case the pair of electrons would carry the fixed energy of the decay—just as the electron in single-beta decay would have had a fixed energy if no neutrino were emitted.

If neutrinoless double-beta decay were to be observed, it would show that the neutrino is, at least in part, its own antiparticle, because it would have to be emitted by one beta-decaying neutron as a particle and absorbed by the other as an antiparticle. This would make it a fundamentally different kind of object from all the other particles of the Standard Model. That could lead to an understanding of why neutrino masses are so small. It could also provide still another source of violation of the combined charge-conjugation and parity symmetry, helping to explain why the universe contains so much more matter than antimatter. It would certainly give some important pointers as to what lies in the far east, off the edge of our current map.

The challenge for experiments searching for rare decays is to eliminate, as much as possible, the noise from natural background radiation. Every component has to be screened for traces of natural radioactivity (mostly uranium and thorium) using specialized instruments. The detector construction must be carried out in a carefully controlled, clean environment to avoid any contamination during assembly. Unsurprisingly building such an instrument takes a long time. The experiments then have to be installed far underground, and watched carefully for a long time. Several groups are building and running experiments right now.

We have learned much more than one might have expected from exploring the Neutrino Sector: It forced a significant change in the Standard Model, it holds out a possible new source of matter-antimatter asymmetry, and it may have given us some important clues as to what is happening far in the east, at energies beyond our current reach. The voyages of exploration and discovery continue, deep into the Neutrino Sector. It may not be wilderness for long. But what lies in store for our band of explorers right now is a long-haul flight back east to the airline hub. The land of Bosonia, home of the W, the Z, the photon, and the gluon, requires our attention. And a mystery lies at its heart.

EXPEDITION VII

Into Bosonia

The benefits of airport coffee – Thoughts on a frozen windshield – Volcanoes and how to avoid them – Carnivorous particles – A triumph and a welcome

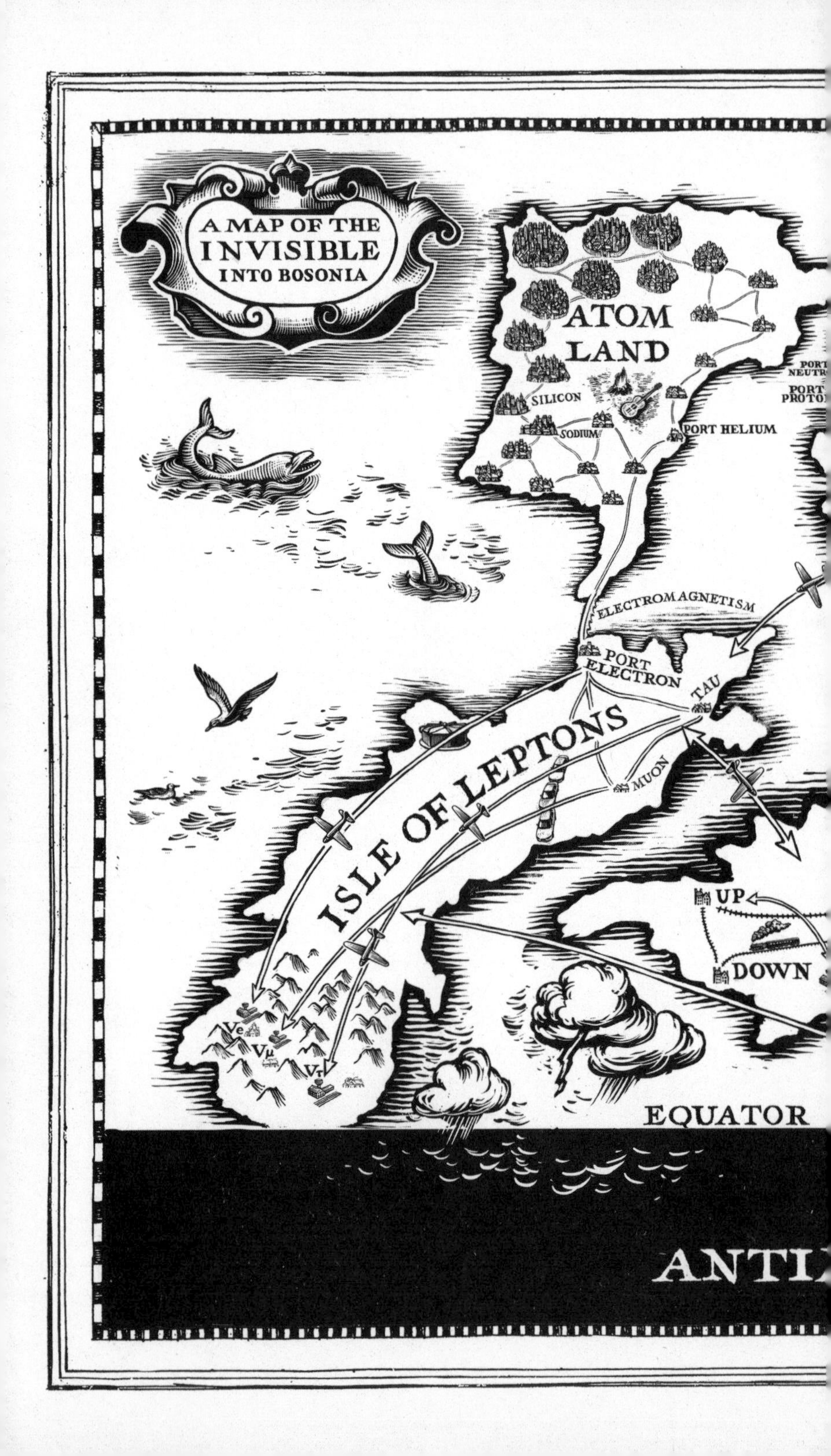

A MAP OF THE
INVISIBLE
INTO BOSONIA
ATOM
LAND
SILICON
SODIUM
PORT HELIUM
ELECTROMAGNETISM
PORT
ELECTRON
TAU
ISLE OF LEPTONS
MUON
UP
DOWN
Ve
Vμ
Vτ
EQUATOR

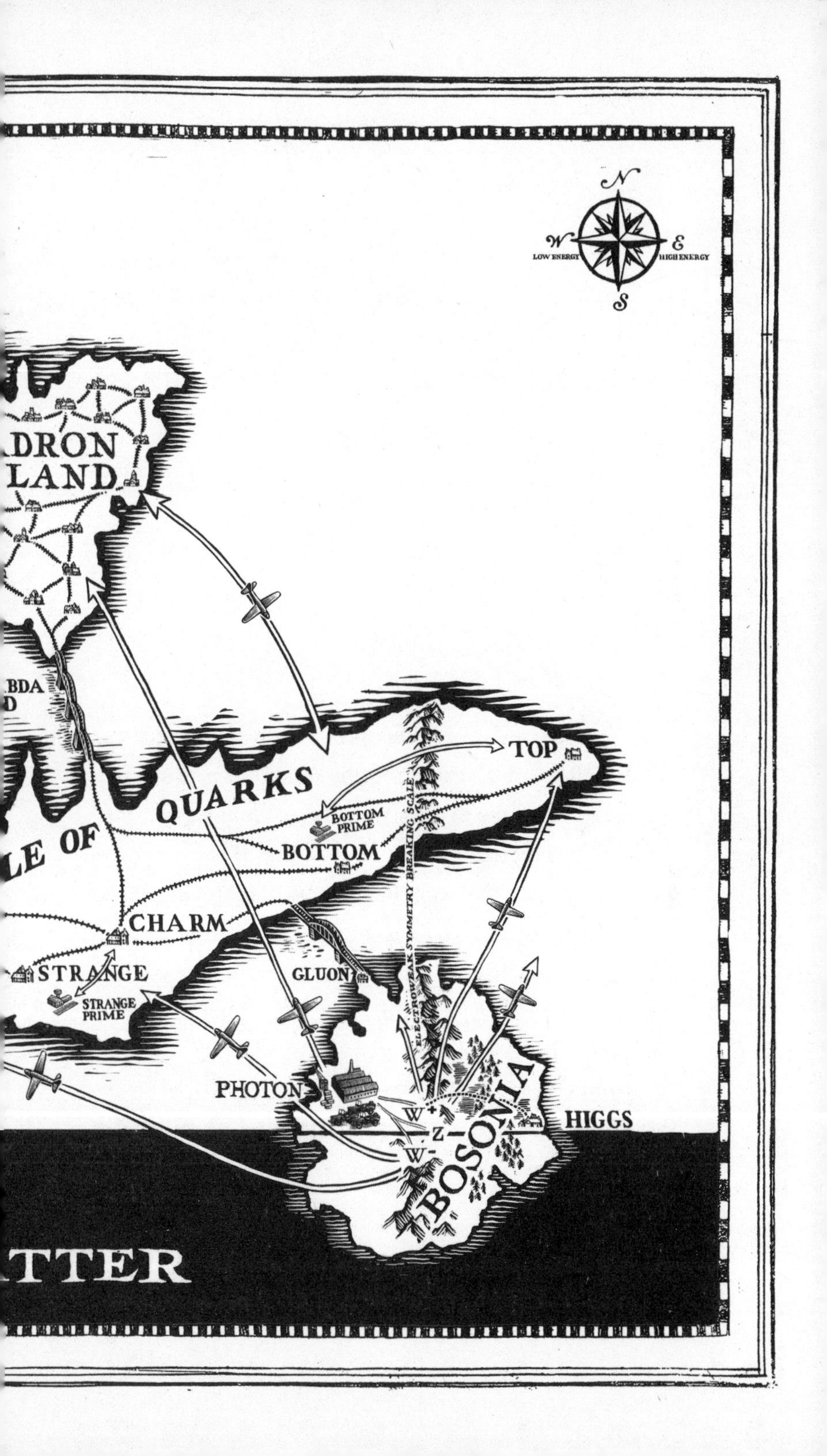
N
W
E
S
LOW ENERGY
HIGH ENERGY
DRON
LAND
QUARKS
LE OF
TOP
BOTTOM PRIME
BOTTOM
CHARM
STRANGE
STRANGE PRIME
GLUON
ELECTROWEAK SYMMETRY BREAKING SCALE
PHOTON
W+
Z
W-
BOSONIA
HIGGS
TTER

26

Symmetry and Conservation

A restless red-eye flight and a bumpy landing later, and we are ready to embark upon a thorough exploration of the last known major landmass on our map—the enigmatic land of Bosonia. This is where the W, the Z, the gluon, and the photon are all based. It is where the cars, trains, and airplanes that connect the other islands originate—all the forces of the Standard Model. And in the heart of its interior lies an enigma. This is our next goal.

Bosons are different. All the matter particles we have encountered—the electron, muon and tau, the quarks, even the neutrinos—have been fermions, meaning that they carry a half-integer unit of spin.

Bosons carry an integer unit, and this changes their behavior profoundly. Rather than filling up energy levels and excluding other particles—as the electrons do around atoms, for example, in line with the exclusion principle—bosons gang up and are more than content to crowd into the same space. This is just one way in which Bosonia is a peculiar place. Before striking inland, we need to properly equip ourselves. The woman at the information desk at the airport informs us, in no uncertain terms, that one of the important things to take with us is a better understanding of symmetry. Understanding symmetry, she assures us, will actually allow us to see where the forces of the Standard Model, and hence the bosons, come from. She hands us an explanatory leaflet.

We have stumbled across symmetry several times already, in electromagnetism, in relativity, in the discovery of antimatter and the ideas of parity and charge-conjugation. It is a crucial idea in physics; it lies behind basic laws such as the conservation of energy, and behind the forces of the Standard Model, as well as gravity. So we believe what the woman at the desk told us: We need to make sure we know what we are doing before we set off. Otherwise we may lose ourselves in the wild terrain inland. We retire to a nearby coffee bar to read carefully the leaflet we were given. It begins rather philosophically.

The reason that symmetry is so powerful in solving problems in physics is that it is connected to the attempt to be objective. In some sense, the use of symmetry allows us to ask whether things would be the same if we were looking from another point of view.

To say that some physical system respects "translational symmetry" is to state that the laws of physics are the same wherever you happen to be. Rotational symmetry means the angle you look from doesn't matter. As we will see, these symmetries are connected to fundamental physical laws. But it is also more metaphorically true. What would physics look like if we grew up on a different planet? Or if we were all made of antimatter instead of matter? Or . . . ? If the physical laws that we discover really do describe the objective nature of the universe, then they ought to be the same, whatever our subjective point of view and wherever we travel, on our allegorical map or on a real one.

If some physical object, or an equation describing some physics, is symmetric under some hypothetical change, that means it looks the same after the change as it did before. So the change is, in some sense, no change at all. For example, a circle is symmetric with respect to rotations around its center—it looks the same no matter what angle you rotate it through.

A square is also symmetric, but only under quarter rotations—that is, rotations through 90, 180, 270, or 360 degrees. For any other angle the sides will have changed direction; they won't line up with the original. So the perfect rotational symmetry of the circle is broken.

This idea can be generalized to other transformations, and we have already found some of these on our journeys. Parity symmetry is the idea that the world is symmetric under reflection—that reflecting everything in a mirror would leave physics just the same. We know that this is not actually true, because the weak force behaves differently for particles of different handedness. Likewise, charge-conjugation symmetry would hold if the transformation of all particles into antiparticles and vice versa left physics the same. Again, the weak force messes this up, as well as in a small way breaking the combined transformation of charge-conjugation and parity at the same time. It is as though the circle has an imperfection in the rim, a glitch, that allows us tell when it has been turned.

The major impact of symmetries on the laws of physics is encapsulated in a theorem proved by the German mathematician Emmy Noether in 1915. Omitting technical qualifiers, Noether's theorem states that for every continuous symmetry in nature, there exists a conserved quantity, something that can neither be created nor destroyed.

If some physical system has a "continuous symmetry," that means that there is a variable that can be changed to any value without affecting the system. The rotation of a circle is an example—any angle of rotation is fine, so the symmetry is continuous. For a square, only four angles are valid, so the symmetry is discrete instead. An example of a continuous symmetry that we think applies to the laws of physics is "translational symmetry": the fact that the laws of physics don't depend on where you are—they are the same everywhere. This is a continuous symmetry with respect to changes of location. Noether's theorem says that because of this, there must be a conserved quantity.

The conserved quantity in this case is momentum; we have seen the conservation of momentum in action already, but now we can see where it comes from. From an assumption that the laws of physics are the same wherever you are, Noether's theorem allows us to prove the conservation of momentum, and thus start deriving the laws of motion. This is a very powerful tool in our explorer's toolbox, and we are going to be applying it to the photon, our first stopping point in Bosonia.

There is a continuous symmetry in electromagnetism. It is apparent in the fact that electric current flows from high voltages to lower voltages, but only voltage differences matter. The absolute voltage has no meaning. This is why birds can sit on high-voltage electric cables without turning into tasty fried snacks. The wires are at a high voltage, but as long as the birds are at the same voltage as the wires, no electric current flows and no harm is done.

This means that if we could, hypothetically, change all the voltages everywhere in the world at once, it would make no difference to anything. In fact for all you know, it just happened then, while you were reading that sentence. Invariance under changes of voltage is a continuous symmetry of the equations. In QED, the quantum theory of electromagnetism, the electron is a wave-like quantum state, and the voltage symmetry is now manifest as the fact that physics is invariant when the phase of this quantum state is changed. Remember, the phase just specifies the absolute positions of the peaks and troughs of the associated wave. Only differences in phase matter—the value of the phase itself, just like the value of voltage, makes no difference to anything. Noether's theorem tells us that there should be a conserved quantity corresponding to this symmetry. There is. The conserved quantity is electric charge, which we already saw from Maxwell's equations cannot be created or destroyed.

The field of mathematics that deals with symmetries is called "group theory." A "group" in the mathematical sense is the list of all the different possible actions that satisfy a given symmetry. So the group of possible rotations of a square would have four members: There are four angles you can rotate it through that preserve it unchanged. For the circle, and for any other continuous symmetry, the relevant group has an infinite number of members.

From the point of view of physics, one of the strengths of mathematics is that if you learn some piece of math once, it will very often crop up over and over again in different bits of physics. This happens here, in that the same mathematical symmetry groups crop up in a variety of circumstances, and to help recognize and discuss them, mathematicians have given them names. The relevant group for the phase-invariance symmetry seen in QED is called U(1), where the "1" comes from the fact that the phase is just one number (other groups we meet will have more) and "U" stands for unitary, which is a property that guarantees the conservation of charge, in that if we start with one electron and move it about in the group, we'll always have one electron, albeit with different phases.

We finish our coffees. This was clearly a more important tourist information leaflet than most. The background briefing it has provided, identifying the global U(1) symmetry of QED and its connection with the conservation of electric charge, gets us to the point where we are ready to take on the mission to explore Bosonia. Over to the car rental stand to hire a 4×4. We are heading into some very challenging terrain.

27

Symmetry and Bosons

As we pull out of the airport parking lot and head for the hills of Bosonia, we still clutch the leaflet we got from the information desk. It is satisfying and, the leaflet assures us, important to know that there exists a mathematical proof connecting the fact that charge is conserved with the fact that under a simultaneous phase rotation everywhere in the universe, nothing changes. Conservation of charge is very practical, and is absolutely key to the operation of vast amounts of our technology. The phase of a quantum field seems rather distant and abstract. But mathematics tells us that one implies the other.

So far so good, but it is not at all obvious how this connects with the bosons that carry the fundamental forces, and hence the terrain we are planning to traverse.

That said, the idea of simultaneously changing everything everywhere seems a bit unphysical. We know that instantaneous communication is not possible. In relativity, the very idea of "simultaneous" events is subjective, and depends upon the speed of the observer. People traveling at different relative speeds will disagree on whether or not events occur at the same time.

Maybe we should insist that physics remains the same under all changes of phase,[48] even if there are different changes in different places? This would be called a local symmetry, whereas before we had a global one.

If we do enforce this symmetry locally, allowing ourselves to make different changes to the phase of electrons at different points in space, something marvelous happens. The need to retain the local symmetry means that extra terms have to be added in the equations that describe the electron. This is a bit like the terrain we are driving across in our Jeep. The driver, who is slightly obsessive, wants to maintain a constant speed so she knows we will keep to our demanding schedule. To keep a constant speed on a plain or plateau, she needs the same amount of power from the engine.

48. That is, rotations in the U(1) group.

Whether we are on the coastal plane or a high plateau is irrelevant. Indeed, it wouldn't matter if someone mysteriously raised Bosonia a hundred meters higher above sea level. So long as they raised the whole country at once, a plain or plateau would remain flat, and the Jeep would stay at the same speed without the driver needing to move her foot on the accelerator. However, if there are local changes in height—hills—then to maintain constant speed, the force exerted by the engine has to be changed. If we are going up, it has to increase, and going down, it needs to decrease.

The need to change the power to the engine is the analogue of the extra terms we need to add to the equations of motion of the electron. These terms look somewhat clunky and awkward at first sight, but inspecting them properly, and seeing what effect they have, reveals that they are *exactly the quantum mechanical version of Maxwell's equations*! This is phenomenal. We enforce a local U(1) symmetry, and in doing so we produce the equations that describe the photon, and the whole of QED, as if from nowhere! Just as the driver has to use force to keep the speed constant when the slope varies, QED appears as a force in the Standard Model when the phase is allowed to vary locally.

This startling success makes some sense in retrospect. We know that the phase symmetry is connected to conservation of charge, and we know that local differences in the amount of electric charge lead to electromagnetic forces. But even so, to derive the existence of the photon from a symmetry this way is beautiful and remarkable, and is a key insight into the nature of Bosonia.

The photon is the first boson we heard about on our travels—Photon is a major industrial city on the west coast of Bosonia. It manufactures the cars traveling the road networks in the Isles of Quarks and Leptons, Hadron Island, and Atom Land.

It seems clear now why the woman at the information desk was so emphatic that to properly explore Bosonia, we needed to be tooled up with a decent knowledge of symmetry. The whole of the road network and the Photon industrial zone can be traced back to a local U(1) phase symmetry. It is tempting to wonder whether a trick as good as that can be tried again. There are symmetry groups other than U(1) available. What might they do for us?

They can do a lot, it turns out. There is a group known as SU(2), which instead of the one phase of U(1) has a little vector of two numbers.[49] Enforcing local invariance under this symmetry group, just like we did with U(1), gives us something very much like the W and Z bosons of the weak force; we are closer to understanding our airlines. And enforcing local symmetries

49. The "S" means it is a special subgroup of a bigger U(2) group.

for the group known as SU(3) produces the whole of the strong force, the railway networks of Hadron Island and the Isle of Quarks and the connecting bridge to the gluon on the northwest cost of Bosonia. This means that all the forces of the Standard Model are related to local invariance under these symmetry groups.

Symmetry is a beautiful and pleasing feature of nature, seen in the growth of plants, the arc of a rainbow, and the flow of the seasons. But the close relationship between symmetry and the forces of the Standard Model turns out to be more than just an aesthetically pleasing coincidence; it seems to be essential.

28

Virtual Particles and the Defense Against Infinity

Leaving Photon in our rearview mirror now, we turn and head east to continue our journey into central Bosonia. The roads are getting rougher, and the geology has a rugged and unstable look to it. There have been stories of volcanic activity in the region, and the occasional geyser seems to lend them credibility. Since Bosonia is where all the forces originate, perhaps we should expect turbulence.

As we have seen several times on our previous expeditions, an interaction between two particles in the Standard Model is described by an exchange of bosons. Electromagnetic attraction and repulsion involve the exchange of photons (our road network); the strong interaction is mediated by the exchange of gluons (railways), and the weak force by the W and the Z (those airlines, with their sometimes oddly placed airports).

Part of the power of the Standard Model is that it can provide you with a quick estimate of the prediction for some process—say, for the chances of two particles scattering off each other. And if you need a more precise answer you can do more work, more calculations, and gradually reduce the uncertainties in the prediction.

The rough answer for the probability of two electrons scattering off each other could be calculated by considering the chances that they exchange a single photon, because this is the simplest, most likely way such a scatter could occur. But if we want to make the answer more accurate, we should also consider the chances that two photons are exchanged. This is less likely to happen, because each time an electron and a photon interact, we introduce a factor of about 1/137 (0.007 or so) into the probability.[50] But if we want a precise answer we need to include it. Remember that all possibilities have to be taken into account in quantum mechanics, so if we miss some, we get the wrong answer. The more we include, the more accurate our predictions will be.

50. This number characterizes the strength of the interaction, and depends on the charge of the electron—if the charge were higher, the number would be higher.

We could also include the chance of exchanging three, four, or more photons. Each time we increase the number of photons, we decrease the size of the contribution to the probability, making smaller and smaller corrections and converging on a more precise answer. This idea of starting with a rough answer and systematically improving it with more sophisticated corrections is called perturbation theory.

The exchanged particles in all these cases are what we call "virtual." They are very ephemeral and never directly observed. In fact we have met virtual particles before, going around the loops that affect the magnetic moments of the electron and the muon, and bubbling in the quantum vacuum while quarks turn into hadrons. Virtual particles are real in some sense, because if they are not included in our calculations, the answers come out wrong. And they have some relation to real particles. For example, the mass of a virtual particle does not have to be the same as the mass the particle would have if it were real. Virtual photons can have non-zero mass, while real ones are always massless. But the further the mass of a virtual photon is from zero, the less likely it is to be exchanged, and the lower the probability of the scattering. So there is a close connection between the real and virtual versions of the photon, or indeed the real and virtual versions of any particle. In the end, all that actually registers in a detector are the real particles that go into a scatter and the real particles that come out, but the properties of the virtual particles critically influence where they go, and how often.

There are further implications of the virtual particle exchanges, though, as we look closer at the entities on our map. According to the Standard Model, virtual particles can also form little loops, connecting back on themselves, or splitting into little particle-antiparticle pairs before re-forming. In any quantum system this kind of unruly behavior is going on all the time, if you look closely enough. As we study it more carefully, we see that even an electron, on its own, is surrounded by a cloud of virtual particles, looping out and around and connecting back again to the electron. This is what we saw when looking at the quantum corrections to g and the magnetic moments of the electron and muon that we observed on our tour of the Isle of Leptons.

All this may seem quite wild and abstract, but it works. The incredibly precise calculations of the magnetic moments of the electron and muon, discussed when we were exploring the Isle of Leptons, require virtual-particle loops. The existence and mass of the top quark were predicted before its discovery, through the influence of its appearance in quantum loops contributing to precisely measured quantities in electron-proton annihilation.

Since the particles going around these loops carry energy, and since energy is mass multiplied by the speed of light squared, the loops will contribute to the mass we measure for a particle, even if the loops are too small and quick for us to see. It is tempting to hope that it might be possible to calculate the mass of a particle from the energy in all the possible loops that surround it. Maybe that can even explain the curious distribution of the masses of leptons and quarks, scattered from west to east on our map, from the super-light neutrinos to the top quark with a mass nearly 200 times that of the proton?

Sadly, no. Even worse, not only do we fail to get the right answers for the masses, we don't even get sensible answers. The result comes out as infinity. It's as though we are peering closely at one of the bosons by the roadside and a volcano suddenly erupts!

This is not just wrong, it is silly. And a similar thing happens if we try to calculate how the loops affect the charge of the electron. This is bad news. If our fancy theory is predicting everything to be infinity, this is against experimental evidence, to put it mildly.

And this is where the fact that the forces involved in these quantum loops are based on symmetries comes back into play, to save us.

We need to pull to the side of the road and pause for a moment. This talk of infinities and sudden volcanoes is unnerving.

When confronted with an answer like infinity for the electron mass, a physicist could just take the view, "Well that's obviously nonsense, but I know what the answer should be really, because I have measured it." We could take the equations behind the theory and, in all the places where the electron mass appears, with its infinite loop contributions, cross it out and replace it with the actual measured value. This is a reasonable, pragmatic approach, even if it does feel a bit like cheating. It is called "renormalization," because the infinities are being normalized into something sensible. But even as a cheat, this falls flat on its face, in general.

The problem with it is this idea that we can improve the precision of a calculation by adding more loops and exchanges. So we get an infinite answer when calculating all the possibilities with one loop in them, then we replace that by the measured mass. Then we try to improve our accuracy by calculating the next set of loops. Infinity again. Which means we have to fix it to the measurement again, and our new calculation hasn't helped at all. This happens every time we try to improve our calculation. It is frustrating, and it means that in the end the theory can't make proper predictions. Renormalization does not work, in general.

However, there is a class of theories for which normalization *does* work, and has been proved to work. If the theory is based on a local symmetry, then replacing the infinities by the measured values *just once* is all we need. The symmetries guarantee that the infinite bits of the quantum corrections cancel each other out, leaving the finite bits that do indeed improve the accuracy of the calculation.[51] After that, no more of these impossible infinities crop up, no matter how many loops we include.

To our immense relief, all the forces of the Standard Model are constructed this way, with electromagnetism based on the U(1) symmetry, the weak force on SU(2), and the strong force on SU(3). For these theories, renormalization works. We cheat once, and from then on the game is fair. This is probably not a coincidence. Maybe building a theory on symmetry is the only way to make one that can withstand these infinities.

This is massive progress, and we start the engine, pull back onto the road, and continue on our way, somewhat reassured.

51. The proof that this is the case was published by Dutch physicists Gerardus 't Hooft and Martinus Veltman in 1971.

29

Mass and Hidden Symmetry

Our Jeep bounces along over the rock-strewn road, and the countryside gets wilder and stranger with every mile as we continue east. We are less worried than we were that we might be abruptly immolated in an infinity of lava, but there is still something amiss in the interior of Bosonia. And as so many times before, it is the weak force that is responsible for the weirdness.

The group SU(2) gives us "something like" the W and Z bosons of the weak force. But not exactly like them. It gives us a W and a Z with zero mass. This is a big problem. We landed here on Bosonia at the W and the Z. We know they have mass. But giving mass to bosons breaks the symmetry, and that threatens to unleash infinities again.

This is not a problem for the strong force. The gluon is massless, so the SU(3) symmetry can hold, renormalization works, and the infinities in the loops can be vanquished. Likewise for electromagnetism, the photon has zero mass, so U(1) is fine.

The W and Z bosons, however, are both very massive. They are dozens of times more massive than the proton. Those masses are not a minor matter. They are the very reason the weak force is weak and short-range. Plus, W and Z bosons have been produced in particle colliders, and their masses have been measured directly, so there is no doubt about them. They lie to the east of the photon and gluon. But the W and Z masses break the SU(2) symmetry, which on the face of it means that renormalization does not work, the infinities are rampant, and the theory is useless.

Luckily, there are ways of hiding symmetries. It happens quite a lot in nature, not just in the esoteric reaches of the east of our map. For example, think of the formation of a crystal from a gas or a liquid. The gas or liquid is featureless—it looks the same from any angle, and from any position inside the liquid.

For several miles we have been climbing a steep mountain road and driving through a downpour. Rain batters the roof and hood of our Jeep. On the front windshield, the wipers are sweeping frantically, and the side window, as we peer out in an attempt to see where we are, is basically a sheet of water. As we continue to climb, the clouds thin, the rain eases, and the temperature drops dramatically. The water on the windshield begins to freeze as we watch.

Imagine a tiny physicist on the windshield, inside the sheet of liquid, looking around. They are so tiny that the window appears as an infinite surface with a deep pool of water covering it. They can't tell where they are, or which direction they are facing; it all looks the same.

Then the liquid condenses and starts to form crystals. Delicate fern leaves of ice spread across the window. Now, not all directions are the same. Some are parallel to the crystal fronds of the fern, and some are not. The tiny physics gnome can see all the water molecules lining up in planes and rows and can define different angles relative to those planes. Similarly, while the gnome still cannot say where exactly they are on the window, they can certainly say whether they are on a plane of water molecules in the crystal or between planes. Not all positions are the same anymore.

Being a physicist, the gnome will know the importance of this, as should we. A symmetry that was present in the liquid, and which is in fact still present in the underlying theory governing the interactions between the water molecules (electromagnetism again), is not present in the crystal. It has been hidden. The liquid water is in a higher-energy state than the ice crystals, and the symmetry is hidden because the orderly crystal is a lower-energy state. As they lose energy to their surroundings, the atoms arrange themselves into positions that minimize their total energy. That energy is made up of their energy of motion (the slower they move, the less kinetic energy they have) and their potential energy, which they have because of the forces between them. The result is condensation into a crystal, which, while still quite symmetric and rather pretty, has less symmetry than the initial sheet of water.

The second law of thermodynamics, one of those general principles that seem to apply right across our map and beyond, states that entropy always increases. Entropy is a quantity that measures disorder in some sense. Entropy increases when a particle decays, for example, because a unique distribution of energy—all of it in one particle—converts into several particles in which energy can be distributed in several different ways. The result is a less unique, more disordered state. This is why, as we saw earlier in our travels, particles will always decay if they can—it increases entropy.

Given all this, one might worry about how a more orderly state can spontaneously arise from a less orderly one, as seems to happen in crystallization. However, the redistribution of the energy to the environment as the liquid cools means that the overall state (including both the crystal and its surroundings) has higher entropy than the starting liquid. For the same reason, evolution from primeval sludge to Homo sapiens does not violate the second law of thermodynamics, once you take into account the fact that there is a nearby thermonuclear reactor (the Sun) madly redistributing energy that can drive the process.

Both crystallization and evolution are vastly complex and difficult to predict. Physics is easier—all we need to know is that cooling and redistribution of energy can hide symmetries. Even if we have a symmetrical theory for the interaction between particles, it is still possible that those particles form low-energy states that do not show all the same symmetries. This is a pointer toward what is going on in the middle of Bosonia, and the solution to the problem of W and Z boson mass.

This idea of *hidden* symmetry is just as important as the idea of symmetry, which gives us conservation laws and keeps infinities at bay. It is fundamental to physics that there are symmetries, present in the underlying laws governing the behavior of particles and forces, that are hidden in low-energy states, meaning that in general they are not apparent in everyday life.

Remember, we had a problem with the bosons of the weak force. To protect against infinities, we need a symmetry to be present—the group SU(2), in this case. But we also know the W and Z bosons have mass, which breaks that symmetry. Freezing and crystallization point to a way out of this conundrum. Freezing a liquid into a crystal hides the rotational and translation symmetries inherent in electromagnetism by arranging the atoms or molecules of a material into a lower-energy state that happens to have more order and structure than the higher-energy liquid state. In a very similar way, the Standard Model hides the symmetries inherent in the weak force. The energy at which this happens is a very significant longitude on our map, and it is this scale we have now encountered as we top the ridge of the mountains we have been climbing and gaze out across the interior forests and prairies of central Bosonia, right across to the eastern coast.

As far as the W and the Z, and the rest of the Standard Model, are concerned, "everyday life" corresponds to energies below this scale, behind us and to the west on the map. Back there, the SU(2) symmetry of the weak force was hidden. Beyond this point, things change. We have reached the "electroweak symmetry breaking scale."

30
Electroweak Symmetry Breaking

The ridge of mountains marking the Electroweak Symmetry Breaking Scale follows a rift between continental plates that runs north-south across all latitudes of our map, even into the southern hemisphere. As we start to descend from the pass we have taken into the forests below us, we expect surprises.

The significant changes in physics at this longitude affect many experiments and measurements. Perhaps the clearest is in the process that we last encountered on our train journeys around Hadron Island and the Isle of Quarks—the scattering of electrons off protons, which first revealed the presence of quarks inside hadrons.

In these experiments, a beam of leptons—usually electrons or positrons—scatters off a hadron—usually a proton—and transfers so much energy and momentum that the hadron shatters. Usually this is an electromagnetic interaction, meaning that a virtual photon is exchanged between the electron and a quark.

However, electron-proton collisions can also be mediated by the weak force, meaning the exchange of the W or the Z boson. Since the W carries away the electric charge of the electron, in those events the emerging scattered lepton is a neutrino.[52] As might be expected, these "charged-current" events are much rarer than the electromagnetic scatters, reflecting the fact that the weak force is much . . . well, weaker than the electromagnetic force.

As the energy scale increases and we move eastward, the rates of these two different types of scattering converge until, as we descend, they become roughly equal. As we descend the ridge of the Electroweak Symmetry Breaking Scale, they become equal. And, if we can pull together what we have learned on our journeys so far, we already know enough to understand why.

In the west, at lower energies, most of the difference in strength and range between electromagnetism and the weak force is due to the different masses of the

52. This is the W boson playing its trick of switching particles around within doublets, which we noted when we first studied the weak force.

exchanged bosons. The photon has zero mass, while the W and Z have masses of just below 10^{11} electron volts—about a hundred times the mass of a hydrogen atom.

The range and strength of the force depend on the mass of the exchanged particle because the particles in these exchanges are virtual and do not have the correct mass. In fact they *cannot* have the correct mass—there is no way that an electron can suddenly emit a real photon, a W or a Z, and conserve energy. The only way that can work is if the emitted particle has the wrong mass. And we already know that the further the virtual particle is from the correct mass, the lower the probability of it being emitted or exchanged.

In the low-energy interactions of Atom Land, and indeed everyday life westward, the massive W and Z bosons are much further from their correct masses than is the photon, so they are much less likely to feature. This is why the electromagnetic force is stronger and more commonly observed.

Go farther east, though, where energies are higher, and the difference in mass between the W, Z, and photon becomes less and less relevant, and the weak force and the electromagnetic force become very similar in strength.[53]

53. The forces don't exactly unify (they still come from distinct symmetries, U(1) and SU(2)) but it turns out there is a mixing between the neutral bosons. The photon is not exactly the neutral boson you would expect from U(1), and the Z is not purely from SU(2). They are both a mixture of the neutral bosons one would expect from the two different symmetries.

In terms of the transportation network on our map, beyond this ridge of mountains flying has become just as common as traveling by road, mainly because the roads are becoming more difficult to negotiate.

This makes the mass problem even more important. The W and Z masses play an absolutely crucial role in determining the characteristics of the weak force, and we still have the problem that they ought to break renormalization and let in the infinities. As indicated by clues on our frozen windshield, and the name of the Electroweak Symmetry Breaking Scale mountains we have just passed through, a broken symmetry is behind this.

People suspected this might be the case decades ago, before the rest of the Standard Model was established and even before the Isle of Quarks was discovered. If fundamental particles—whatever they might turn out to be—were to have mass, something had to be done. Something strange and different has to be going on in the forests we are now entering.

31

Hunting the Higgs

At the bottom of the mountain road, we make camp and a small party sets off to explore on foot, carefully, to study the flora and fauna of this new land. Peering from behind some dense vegetation, a compelling and dramatic sight meets our eyes: Some large predators are stalking some even larger grazing animals. One of our guides recognizes the species and narrates the scene in a hushed voice:

The weak bosons have formed a small hunting pack.

Silently, stealthily, stalking through the undergrowth, they approach the unsuspecting Goldstone bosons.

One of the Goldstone bosons looks up suddenly, perhaps sensing something.

The weak bosons pounce! Three of the Goldstone bosons are brought down, messily devoured, and digested. They will provide much-needed longitudinal polarization states for the weak bosons. But the fourth, more alert Goldstone boson escapes! Acquiring mass of its own, the spinless particle flees eastward to safety and hides, to live another day . . .

Clearly Bosonia is a brutal place, but that piece of natural history contains the solution to the problem of the W and Z masses. The solution is wide-ranging and requires a few steps, a few additions to the Standard Model of physics that we have been exploring.

To try to build a theory that can accommodate all this, the first step is a dramatic one. We must invent a new quantum field permeating the whole of our map, the whole of the universe, unlike any other field we have encountered so far. This field carries no spin and no charge.

For the second step, this field has to be such that it has a hidden symmetry, similar to the frozen water on our windshield up in the mountains, but with one extra twist. Imagine the water molecules all carry a magnetic dipole. In the liquid, they will all point in different directions and be continually knocking each other into different alignments as they jiggle around enthusiastically. There is no overall magnetic field. As the water cools and freezes, the jiggling lessens, and eventually the molecules will settle down to the extent that all the dipoles point in the same direction, because that is the lowest energy configuration.[54] As this cooling happens, an overall magnetic field develops because of the now-aligned dipoles. This magnetic field breaks the symmetry of the system, too—it chooses a special direction for the north and south poles of the magnetic field, when for the liquid no such direction existed. To deal with the W and Z boson masses, we postulate a quantum field in the universe that mimics this behavior. When the universe is hot and dense, a symmetry is present and the field has an average value of zero. As we reduce the energy, down through the electroweak symmetry breaking scale, the field gets an average value that is not zero.

In the third step, we make the mass of the fundamental particles a property they acquire by interaction with this non-zero value of the field. This can be arranged mathematically for a field like this (with no spin and no charge). The advantage we achieve by all of this, the goal of these three steps, is that the symmetry is still present in the theory, so the infinities are kept at bay, but the symmetry is absent in everyday life, so the particles can have mass.

54. This is true for types of materials known as ferromagnets.

This is all very neat. A little bit too neat.

When a symmetry is hidden like this, it leaves traces. Specifically, it leaves new ways of transmitting information: new particles, massless bosons that should be seen wandering around Bosonia, that should appear in the quantum loops of the Standard Model, and of which there is no sign anywhere. Precise surveys and measurements leave no room for them on our map. This is a big problem for our neat symmetry breaking trick.

There is even a mathematical theorem, the Goldstone theorem, that says that when a symmetry is hidden, such bosons *must* exist. In the physical examples we have looked at, you can see them. Once a crystal, or a regular array of magnetic dipoles, has been formed, there is a new way of transmitting information through the crystal. Displace an atom slightly from its site in the lattice, or a dipole from its alignment, and the perturbation will ripple through the material. This ripple can transmit energy and information. It is the classical analogue of what, in the quantum field, would be a new massless boson.

On the face of it, the quantum field needed to give mass to the W and the Z gives rise to four such massless bosons, and we have covered enough of the map to know that they aren't there. This is the problem that was being grappled with by several physicists in the

1960s—not specifically for the W and the Z, which weren't known as of then, but generally for the case of massive fundamental particles. In a nutshell the problem is: If massive bosons carry a force, then a symmetry must have been hidden. But if a symmetry is hidden, there should be Goldstone bosons. So where are they?

The answer was discovered by two Belgian physicists, François Englert and Robert Brout, and by Peter Higgs in Edinburgh, and it involves the kind of carnivorous boson food chain we witnessed in action on our first foray into the woods above the Electroweak Symmetry Breaking Scale ridge.

A massless boson that carries a single unit of spin, such as the photon or the gluon, has two possible orientations for its spin. The spin can point along its direction of motion or against it—that defines the helicity, which we met earlier when dealing with very different particles, the neutrinos.

A massless particle has to travel at the speed of light, so the direction of motion is always well defined. There is no reference frame in which the boson is stationary. This would also be true for a massless W or Z. But once they have mass, it no longer works. Where does the spin point if the boson is stationary? The upshot of this problem is that there needs to be an additional option for massive bosons, a so-called "longitudinal" state not present for massless ones, corresponding to the spin pointing perpendicular to the direction of motion when it is moving.

The W and Z bosons appropriate three of the Goldstone bosons to provide these longitudinal states, one each for the W^+ and W^-, and one for the Z. In a sense they devour them. Bosonia is a brutal place, although watching all this happen in the mathematics of the Standard Model is rather beautiful and inspiring, rather like a gory natural history program.

The hunt for that more alert Goldstone boson, the one that got away, has preoccupied much of the last couple of decades of particle physics. It is a scalar particle—meaning it carries no spin, the only such fundamental particle in the Standard Model—and it has now acquired mass. Its existence is predicted by the theoretical mechanism we have put together to explain electroweak symmetry breaking. If we find it in Bosonia, then the weak force and those parts of our map make sense. If not, then we have a big problem. It is the particle we know as the Higgs boson.

Tracking the Higgs through the undergrowth of Bosonia, we had several clues. Precise measurements of the top quark and W masses at the Tevatron proton-antiproton collider near Chicago, combined with even more precise measurements of the Z bosons at CERN and at the Stanford Linear Collider in California, severely limited the territory in which the Higgs boson could be found. Again, if it existed, virtual Higgs bosons should be contributing to already-measured processes via quantum loops.

The Standard Model also predicted exactly how a Higgs boson should appear in the experiment. It would be very short-lived, and would decay to other Standard Model particles at different rates. These rates could be predicted for any given mass of the Higgs boson, but the mass itself was not known and not predicted, beyond the fact that it should be somewhere within reach of the electroweak symmetry breaking scale.

The final push, to cover all the territory on our map that might possibly contain the missing Higgs boson of the Standard Model, came from the Large Hadron Collider (LHC) at CERN. Colliding protons head-on at unprecedented energies, the machine is in effect the most powerful microscope ever built. It gives us access to the whole of Bosonia and beyond. And in 2012 the Higgs was indeed finally observed, first of all decaying to pairs of photons and to pairs of virtual Z bosons, but later in other decay modes, too. Its mass, about 125 giga-electron volts, or about 130 times the mass of the proton, puts it in the eastern reaches of Bosonia, but well within the range required for it to be consistent with the Standard Model.

We emerge from the forests of Bosonia and reach the east coast. Higgs bosons are being studied in great detail in the lively and buzzing maritime city we find here; yet people are still thrilled to hear that we saw a Higgs in its natural habitat. The amazing bosonic ecosystem probably still has things to teach us.

This culmination of our journey was a remarkable triumph, not only for us as explorers, but also for the predictive cartography of the theory. With the knowledge that the Higgs boson exists, we now have a theory that allows the W and Z to have masses as observed, but still contains the symmetries we need to keep infinities at bay. This means the theory is capable of making predictions at energies far above the electroweak scale, far into the unknown east.

EXPEDITION VIII

Far East

Taking stock – Dinner and a decision – The fascination of the east, stories and speculations – Jigsaws and a seamless web – Ships setting sail

A MAP OF THE
INVISIBLE
FAR EAST
ATOM
LAND
SILICON
SODIUM
PORT HELIUM
PORT NEUTRON
PORT PROTON
ELECTROMAGNETISM
PORT
ELECTRON
TAU
LAMBDA
QCD
ISLE OF LEPTONS
MUON
ISLE
UP
DOWN
DOWN PRIME
EQUATOR

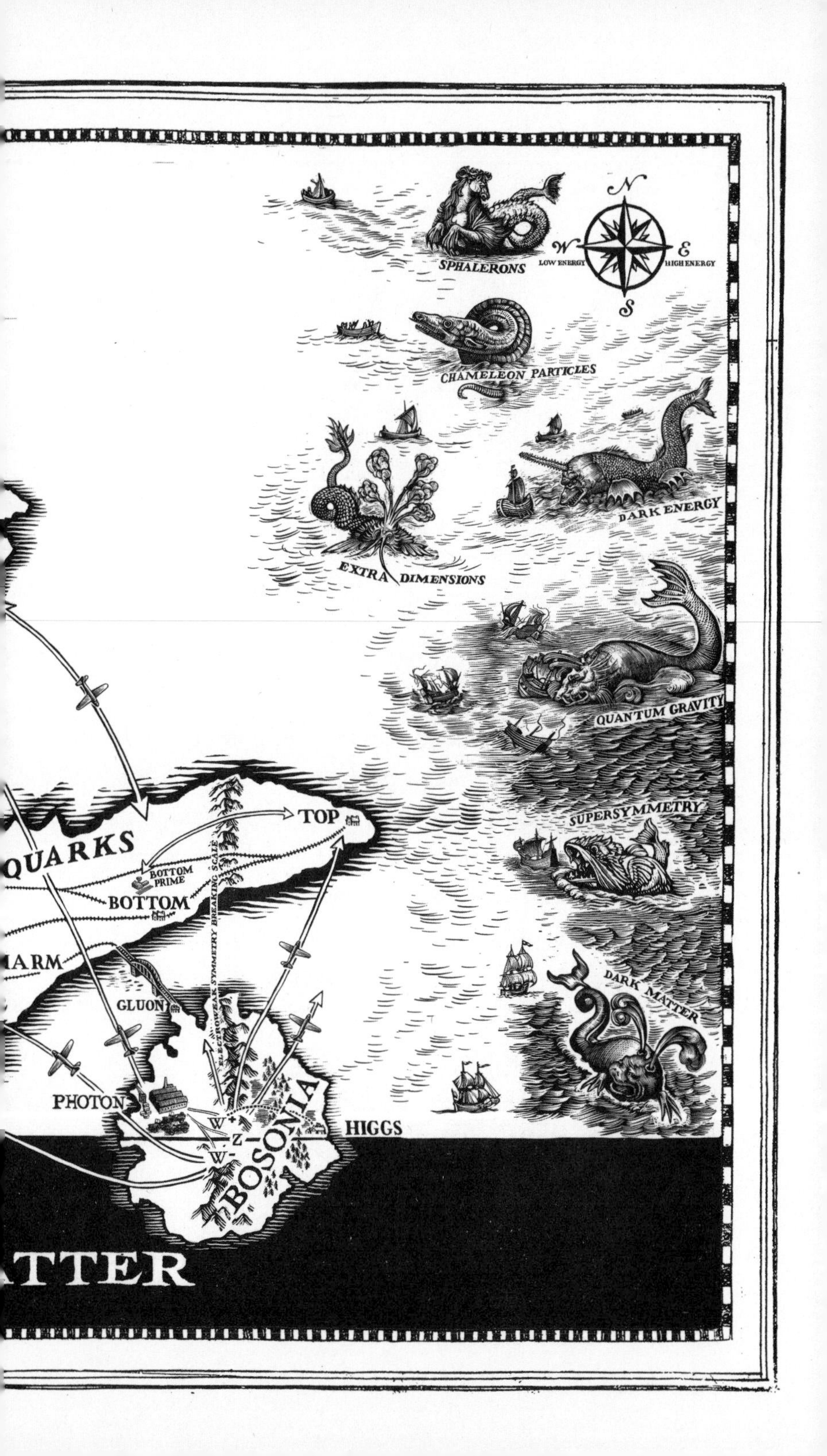
N
W
E
S
LOW ENERGY
HIGH ENERGY
SPHALERONS
CHAMELEON PARTICLES
DARK ENERGY
EXTRA DIMENSIONS
QUANTUM GRAVITY
SUPERSYMMETRY
DARK MATTER
QUARKS
TOP
BOTTOM PRIME
BOTTOM
ARM
GLUON
ELECTROWEAK SYMMETRY BREAKING SCALE
PHOTON
W+
Z
W-
BOSONIA
HIGGS
TTER

32

Why Go?

We stand on the eastern shore of Bosonia, gazing out at the ocean horizon in front of us and thinking back over the lands we have explored in the preceding expeditions. There is some debate. Should we—can we—continue, or is this the end? Should we put our feet up and stay here? There is a rather nice-looking fish restaurant off the promenade, with a terrace overlooking the beach. We find a table for dinner and discuss future plans.

Over the aperitif and starters, we relive our recent travels and discoveries. Predicting, hunting, and finding the Higgs boson was a singular triumph. A problem in the theory had led to the postulation of a qualitatively new and unique object in nature: a fundamental particle with zero spin. Such triumphs occur much more rarely than you might think in physics; Dirac's prediction of antimatter probably matches it, but not much else comes close.

The Standard Model stands tall, then. The different elements of our map all connect and fit together self-consistently.

Atom Land has its orderly array of elements, each containing a nucleus surrounded by electrons, quantum particles bound together by the sophisticated road network of electromagnetism.

Those electrons abide on the Isle of Leptons, along with their two easterly allies, Muon and Tau, also connected by road, plus the Neutrino Sector away down to the south, and west, accessible only by air, the weak force, which, albeit tenuously, connects all the lands we come across.

East of Atom Land, we entered the nucleus and Hadron Island. Protons, neutrons, and the other hadrons are connected by the sophisticated rail network of the strong force, which also takes us into the Isle of Quarks, with its Down, Strange, and the rest, with their primed airports nearby.

And in Bosonia we made sense of the transport networks and hunted down the Higgs near the Electroweak Symmetry Breaking Scale.

There might be a bit of wilderness left over in the Neutrino Sector, but pretty much everything works. Without the discovery of the Higgs, our map would have been just a rough outline, and would definitely have failed around the mountainous Electroweak Symmetry Breaking Scale ridge. With the Higgs, the symmetries hang together, infinities are vanquished, and yet the fundamental particles have mass. With a small number of fundamental objects and principles, the Standard Model describes and predicts an enormous variety of physical phenomena, over energy scales ranging from zero up to several 10^{12} electron volts and potentially far beyond, far into the east of our map. There is a mood among some of our little band that this is enough. Why explore further?

As the main course arrives, others argue back strongly that we should not be too smug and complacent about the power and subtlety of the Standard Model, nor should we despair that there is nothing left to discover. There are some important factors to consider. It is worth remembering that our overall map fails entirely to incorporate one of the fundamental forces. As we saw on our digression, gravity is described by the general theory of relativity, and in that context it is a consequence of the curvature of space-time rather than a force mediated by bosons, like those in the Standard Model. For the regions covered on our map, this is good enough. It is even possible to make a quantum theory of gravity that works in these regions, although it is not renormalizable, which as we saw on our last trip means it is not protected against infinity, so we should expect trouble. Far enough east, at high enough energies, trouble does indeed happen.

Because gravity is not properly included in the Standard Model, if we were to go far enough east, there is definitely a distance scale, corresponding to an energy of around 10^{27} eV, beyond which things definitely go wrong. This is the Planck scale. Gravity at this scale becomes as strong as the other forces. Singularities and infinities break out all over the place, and we have no way of understanding, or predicting, what might happen. This is a very significant failure in the ambition to describe natural phenomena within a single framework. The Standard Model is very clearly not a theory of everything, even if it co-opts general relativity as a partner.

Worse, even at lower energies, the partnership between the Standard Model and general relativity is inadequate.

The amazing breakthrough of gravity is that the same theory describes the way objects fall to the ground on Earth as well as the orbits of the planets and moons in space. This is a classic success of physics—a single set of rules describing a wide range of phenomena. Newton's gravity was a breakthrough, though you need general relativity to get the planetary orbits exactly right (and to predict the timing and orbit of a satellite well enough to make the Global Positioning System work accurately).

As accurate measurements of more distant astrophysical objects—stars, galaxies, clusters of galaxies—are made, the expectation is that our theory of gravity should work for them, too. Specifically, look at the orbit of stars around the centers of galaxies, compared to the orbit of planets around the center of the solar system. For each planetary orbit in the solar system, there is a predicted speed for the orbit. This has to be the case, because the gravitational attraction between the Sun and the planet has to exactly match the required centripetal acceleration of the orbit. The same applies to stars orbiting the center of the galaxy. In the solar system, general relativity gets it right. But for the stars in the galaxy, the answer does not match the observation. The stars are traveling far too fast—their orbital speeds are too high.

The speed of the stars has been measured accurately using the same technique of spectroscopy encountered in our expedition through Atom Land. Atoms give off or absorb light in characteristic patterns dictated by the jumps in energy that their electrons are allowed to make. These patterns allow us to use spectroscopy to identify the elements in stars, as we saw when exploring Atom Land. But we can also observe that sometimes they are shifted, because the star is either receding or approaching. Just as with the classic example of a passing siren, approaching makes the frequency higher—blue shift, for light—and receding makes it lower—red shift, which is more usual, since, on average, stars are receding. Precise measurements of the rotational speeds of galaxies, made by the astronomer Vera Rubin and her team using this technique, convinced people there is a big problem. The stars are going too fast, and the galaxies ought to fly apart.

There are two ways of solving this conundrum. Either our theory of gravity is wrong, or our estimate of the mass of the galaxies is wrong.

If the mass of the galaxies were to be much bigger—by a factor of four or five on average—than the mass of stars and gases that we can see, then the calculations could work again. That would explain the speed of the stars, though the missing mass, if it is there, doesn't seem to be made of any known particle in the Standard Model. As a placeholder, we call it "Dark Matter." Whatever the solution, the problem is a good argument to carry on exploring.

Another gravity-related problem is something currently labeled Dark Energy. From measurements of the red shift and brightness of supernovae, it seems not only that the universe is expanding outward from the Big Bang, but also that the rate of expansion is increasing. Gravity on its own would slow the expansion, as the stars and galaxies attract each other. So a new ingredient is needed to explain the acceleration. We don't know what this new ingredient is, but Dark Energy is the name we give to our ignorance. "Dark" presumably comes from analogy with Dark Matter[55] and "Energy" comes from the fact that it needs to be some form of energy density that is constant throughout all space (unlike matter or photons, which get more spread out as space expands). The quantum loops of the Standard Model actually predict the existence of a vacuum energy like this. Unfortunately, it gets the answer wrong by a factor of 10^{120}—ten followed by 120 zeroes, a number so big I am not even going to try to spell it out in words. It is such an enormous number that even some cosmologists have trouble ignoring it. Again, maybe the answer lies to the east.

The discussion continues as we eat, assembling the possible reasons for exploring farther east. Another reason for further exploration is brought to the table as the desserts arrive. What about matter and antimatter?

55. Which, as US physicist Lisa Randall has pointed out, would be better named "Transparent Matter," really, since it is invisible rather than opaque.

As we explored the Isle of Quarks we discovered that the symmetry between matter and antimatter is broken, meaning that there is a real difference between a world made of matter and one made of antimatter. However, the tiny violations of matter-antimatter symmetry that we know of seem far too small to explain the absolutely gross asymmetry we see around us. Matter is common, and antimatter is extremely rare. Expeditions into the Neutrino Sector of the Isle of Leptons may reveal a source of matter-antimatter asymmetry that could be fitted within the Standard Model. But unless there is something new going on there, beyond the Standard Model, this source of asymmetry also looks too small. An example of "something beyond the Standard Model" that could be discovered there would be that the neutrinos turn out to be their own antiparticles, a possibility we already discussed on that expedition. Such a discovery would also have implications for the far east of our map, meaning there might be supermassive neutrinos to be found out there.

A final, very general argument is brought up over coffee. It is apparent from our explorations so far that, elegant and economical though it is compared to what went before, the Standard Model contains rather a lot of parameters with seemingly arbitrary, but suggestive, values.

For example, the masses of the particles are distributed over a huge range, from the neutrinos in the west to the top quark in the east (not to mention the massless photon, gluon, and possible graviton). Are they really randomly scattered, or is there a pattern in there somewhere? There are some suggestive relationships—for example, the sum of the squares of all the boson masses (Higgs, W, Z) is roughly the same as the sum of the squares of all the fermion masses (quarks and leptons). Is that just a coincidence, or is it a clue?

The Higgs mass is especially troubling. We know all particles get infinite contributions to their masses from quantum loops, and those are removed by inserting the measured mass into the equations, a technique that relies on the symmetry behind the force to protect it from infinities. But with the Higgs boson mass, because of the fact that it has zero spin, the loop corrections are enormous. No infinities, the protection still works, but if you want the Higgs mass to have a reasonable value both at the electroweak scale (which is where it is) and onward into the east to another energy scale, say the Planck scale, the cancellation of those corrections has to be ridiculously "fine-tuned." It is as though the Higgs balances on a knife-edge over many orders of magnitude, where a slip to either side would make nonsense of the Standard Model. A favorite analogy is a bank account with credits and debits (the quantum corrections) of billions of dollars occurring seemingly randomly throughout the month, yet on the last day of each month the account magically contains exactly $125. This would seem to be too much of a coincidence. There must surely be an accountant paying attention—or, in the case of physics, maybe a bigger theory pulling the strings of the Standard Model.

Even the number of generations of matter looks suspicious. Just the first generation would seem to be perfectly adequate to make up all the elements. As we saw, three is the minimum number to allow a real distinction to be made between matter and antimatter, which seems significant. But perhaps we just haven't explored far enough east. Maybe there are four, five, or even an infinite number of generations?

To the relief of the waiters, we pay our bills and leave the restaurant. But the discussion does not die down as we stroll along the seafront back toward our hotel.

33

Clues and Constraints

The comment about extra generations of matter, made as we left the restaurant, leads to a robust response. While it is not easy to rule out anything if it is far enough eastward, we are not in complete ignorance either, even if we have not visited it. We do know that any possible further generations of matter cannot be very similar to the first three—specifically, we know that they do not contain low-mass neutrinos, as the other three do.

We know this from our detailed explorations of Bosonia, especially studies of the Z boson, produced in electron-positron collisions at the Large Electron-Positron Collider at CERN and the Stanford Linear Collider in California.

Those experiments produced the Z boson in the annihilation of electrons and positrons. The Z boson will decay very rapidly into less massive particles, so we don't detect it directly; it is one of those virtual particles that do not have to have exactly the right mass. However, a real Z would have a mass of 91 GeV, and if you tune up your beam energies to 45.5 GeV each, the center-of-mass energy is just right to produce real Z bosons. This is what the experiments did, and at that value, there is a big peak in the probability of the electron-positron annihilation, just because the Z exists.

Still, the Z can be virtual, too, in which case it does not have to have exactly the right mass, and if the beam energies are moved gradually away from the optimal value, the probability drops relatively slowly—it does not fall instantly to zero. Virtual Z bosons can still be exchanged. The rate at which the probability drops on either side of the peak defines what we call the "decay width" of the Z.

The decay width can be thought of as an uncertainty in mass, or energy. Heisenberg's uncertainty principle is reasonably well-known for stating that we cannot know both the position and the momentum of a particle with absolute certainty. The more precisely we know the momentum, the less certain we can be of the position, and vice versa. The same principle applies to time and energy. The uncertainty in energy is related to the uncertainty in time. If the uncertainty in time is short, the energy cannot be precisely known, whereas if the energy is precisely known, the corresponding time is long. What this means for the Z is that if its lifetime is short, its decay width is large. (For stable particles, the lifetime is infinitely long, and the decay width is zero—they have a definite mass.)

How does this tell us anything about the number of generations? It does so because the rate at which the Z boson decays depends on how many particles it can decay to. Sometimes the Z decays to neutrinos, which were invisible to the detectors at CERN and Stanford. But the number of types of neutrino the Z can decay to affects its lifetime, and thus its width. The more neutrinos, the faster it decays, and the broader the width. And the decay width can be measured using visible particles—electrons, muons, or quarks.

It is from this measurement that we know that there are three, and only three, types of neutrino, and thus only three generations. Of course, that is on the assumption that all neutrinos interact in the same way with the Z. There could always be more generations with weird neutrinos; almost anything can happen out east. But any new generations of matter would not really be copies of the three we know of; they would be very different indeed.

More apparently arbitrary features can be seen in the Standard Model. For example, theoretically it would be very easy for the strong interaction to violate the matter-antimatter symmetry, but it does not. Why is that? Is something we don't know about forcing the strong force to respect this symmetry?

It is very tempting to think that these arbitrary features and parameters are fixed inevitably by some so-far-unknown symmetry or principle inherent in a larger theory of which the Standard Model is just a part, in which our whole map is just the western region of a much bigger landscape.

We are tired by now. Many weeks on the move have taken their toll, as has a good dinner and a heated discussion. There are definitely good reasons to think that there are important and interesting discoveries still to be made in the east, but there is no consensus on where exactly to go, what to look for first, or how to travel. We resolve to wait awhile in this rather pleasant seaside town for other explorers to report before we plan new expeditions. We will frequent the harbor bars and pubs and listen to stories. And while we do that, the cartographers in our group will have fun imagining what might be out there. The final expedition of this guide is an expedition of the imagination. What monsters might lie beyond our current map, and what principles, if any, might guide us out there?

34

Sea Monsters and Dark Matters

Theoretical physicists are the imaginative cartographers of physics. They are the fillers-in-of-gaps-with-dragons, the extrapolators of hint and rumor, and the extravagant tale-tellers of the ocean ports. Given the number of open questions the Standard Model leaves them to chew on, it is no surprise that the eastern seas of the imagination are populated with a menagerie of fantastical beasts of varying degrees of plausibility. In this "expedition" we will stay in the pubs along the shoreline and hear what they have to say, in the hope that it may provide guidance for future travels.

The best pub is right on the dockside, on the eastern shores of Bosonia. There are some telescopes of variable quality in the windows, and a lot of bottles behind the dark oak bar. Theorists lounge or slouch on couches and benches, chatting to each other and comparing tales. Occasionally a seafarer, or at least someone who claims to have been to sea, staggers in and is immediately surrounded, plied with drinks, and pumped for any information they have about the lands beyond the horizon.

Many of the stories, dreams, and even (later in the evenings) songs revolve around the fact we discussed over dinner that either our understanding of gravity is wrong, or the majority of the matter in the universe is some unknown form of Dark Matter—probably some particle that lies beyond the Standard Model.

People explore both of those possibilities, but currently the most popular option is the second—Dark Matter. We would dearly like to come across Dark Matter in our journey eastward, and there are a number of ways this could happen.

There is a chance that the Large Hadron Collider (LHC) will manage to produce Dark Matter and (indirectly) observe it. Any observation will have to be indirect, because by its nature the Dark Matter will not interact with the detectors surrounding the points where the protons collide. However, because the detectors surround the collision point, if something energetic and invisible is produced, they will measure a momentum imbalance. That would betray the presence of Dark

Matter, much as the missing momentum in beta decays led Pauli to postulate the existence of the neutrino, all that way back west. Neutrinos are also produced at the LHC, of course, and like Dark Matter they also lead to apparent missing momentum. However, the Standard Model can predict how many of those collision events there should be, and what they should look like. Any anomalies could be a sign of Dark Matter production.

One thing we know about Dark Matter is that if it exists, it interacts gravitationally. Therefore it probably collects near large masses, such as the black holes at the centers of galaxies, or even in the hearts of stars. In these regions of relatively high Dark Matter density, it may be that occasionally two Dark Matter particles meet and, depending on what kind of Dark Matter they are made of, annihilate and produce high-energy photons, neutrinos, or other Standard Model particles. There are telescopes looking for signs of these products of Dark Matter annihilation, including instruments on satellites, where Earth's atmosphere doesn't confuse matters. There are even experiments that use the ice of Antarctica as a neutrino detector—hoping to spot the rare occasions when a very high-energy neutrino collides with water molecules, converting into a muon or an electron and eventually radiating light, radio waves, or other electromagnetic wavelengths.

Sensitive experiments are also built underground, away from the influence of cosmic rays, to search for

Dark Matter interacting directly with the detector. We know that billions of unseen particles pass through us every moment—we are bathed in neutrinos from the Sun, and in low-energy photons left over from the Big Bang. Both types of particles are vital ingredients in our understanding of physics and the universe, and both have been measured—eventually—by highly specialized detectors. If Dark Matter is there, it may also interact with normal matter via the weak force as it passes through Earth—or as Earth passes through the cloud of Dark Matter particles centered on our galaxy.

In this case, there would be a generic kind of Dark Matter particle called a WIMP—Weakly Interacting Massive Particle. This is also the most likely type of Dark Matter to be seen at the LHC. WIMPs present more of a challenge to detect than neutrinos, partly because they are slower, and partly because we don't know what they really are, so (unlike neutrinos) the Standard Model doesn't tell us the interaction probability. So the experiments have to do their best to scan an unknown parameter space, usually mapped out in terms of the WIMP mass and the chances of it interacting with an atomic nucleus.

One of the most sensitive experiments scanning this parameter space to date was called LUX (Large Underground Xenon).[56] LUX was designed to measure

56. As might be expected, it was large, underground (in the Sanford Underground Research Facility in South Dakota), and made mostly of xenon.

both the light and electrons that would be produced if a Dark Matter particle glanced off the nucleus of a xenon atom. It did not see anything, which is a bit of a disappointment. However, more sensitive experiments are under construction. The voyage continues. And the negative or null results are not worthless; they are filling regions of the map that were previously blank. True, they are generally filling in the map with featureless seascape, but even so, at least we know there are no dragons there.

In important ways, the significance of a null result like this depends upon a robust theoretical framework. If there is no theoretical expectation to test, then sometimes a null result really tells you very little, and it may be that you just did a boring experiment. You believed a drunken sailor with a reputation for crazy stories, you left the pub and sailed out there, and you confirmed that he was talking nonsense. Frustrating, especially if it took a lot of time and money.

But with a robust framework—a credible storyteller—a null result can be very important. This was the case when the LHC was searching for the Higgs boson, for example. If the boson had not been found, that would have broken the Standard Model, a very robust theory that has survived for decades and with many precise predictions to its credit. So a null result would in fact have been extremely interesting.

35

Supersymmetry

There is a particularly common and—at least until recently—widely believed seafarers' tale that may be heard everywhere from the most noisome seafront dives to the elegant drawing rooms of the ship-owning aristocracy. This is "supersymmetry." It is a seductive story. It offers many things to the excitable traveler, and one of the things it offers is Dark Matter.

Supersymmetry builds on the extraordinary success, seen throughout our travels, of symmetry principles in physics. It does this by introducing another symmetry, this time between the bosons, which carry the forces, and the fermions, which make up matter. For every spin-one boson, there is a spin-half "superpartner"—a photino for the photon, a gluino for the gluon, a zino for the Z, and the rather unfortunate wino for the W boson.[57] The fermions have spin-zero partners—selectrons, smuons, squarks, and so on.

Supersymmetry cannot be an exact symmetry, because we know that (for example) if a selectron exists, its mass is not the same as the electron mass. In fact nearly all these new particles have to be far enough east—that is, high enough in mass—that we haven't been able to find them yet. Any that are of lower mass have to be deeply hidden in mountains or jungles somewhere else on our map. That means that supersymmetry has to be at least slightly broken, or hidden. But we know that even hidden symmetries can be very important.

One advantage of supersymmetry is that it solves the "fine-tuning" problem with the Higgs mass. Remember the bank account with big credits and debits that always mysteriously canceled out to $125? In the quantum corrections to the Higgs mass, each fermion is a debit and each boson is a credit. Therefore, if you have an equal number of bosons and otherwise identical fermions, cancellation is guaranteed. The account comes to zero. Supersymmetry is the hidden accountant. Even though it has to be hidden, if the masses of the

57. The Higgs, which is always special, has to have at least four extra Higgs bosons, as well as superpartners.

new supersymmetric particles are not too far away east, the cancellation is good enough to solve the problem with the Higgs quantum corrections.

Supersymmetry isn't really a single theory; it is a symmetry that can be used to build many different theories. That's one reason it features in so many of the rumors we hear on the dockside. In many of these theories, supersymmetric particles carry a conserved quantity called R-parity. Conservation of R-parity has the consequence that supersymmetric particles can only decay into other supersymmetric particles, and that in turn means that the lowest mass supersymmetric particle ("sparticle," you guessed it) must be stable. That means lots of them should be left over after the Big Bang, and that makes them good candidates for Dark Matter WIMPs.

Unfortunately, at the time of writing, no sparticles have been found. For several years, supersymmetry theorists have been saying we just need to sail a little farther east and they will show up. The LHC opened up another big area of the map, and there were high hopes that sparticles would be discovered. However, none has been seen so far. Maybe we just need to go still farther east. Or maybe we need to look for other solutions, to listen to other stories.

36

Into Another Dimension?

On cue, another traveler staggers into the bar, shaking the sea spray from her shoulder-length, straw-colored hair, and orders a large gin. She will relate another common yarn, one not quite as respectable as supersymmetry but told by some loud and convincing travelers—the tale of extra dimensions.

This is the idea that space doesn't just have the three dimensions that we can see, but several more, which are accessible only at high energies. Extra dimensions are really odd in that they are fairly easy to describe in mathematics but are pretty much impossible to imagine. In the equations of physics we often count over three dimensions of space, essentially length, breadth, and depth. Momentum has three components, for example, which a physicist would normally label with x, y, and z, or simply 1, 2, and 3. Each component tells us how fast an object is moving in one of those three dimensions. For some purposes we can even count to four dimensions, treating time as the fourth. Mathematically it is easy enough to extend that to count over five, six, or more—just keep counting. From a mathematical point of view, it would seem a bit arbitrary if things only worked with three or four and stopped.

Mathematics deals with abstract concepts and doesn't care about physical reality, of course. Physics cares. From the point of view of physics, what would extra dimensions mean for the natural world? The only way of getting any idea is to reduce the number of dimensions and work by analogy.

So for example, imagine that the whole three-dimensional universe is embedded in some bigger, multidimensional space. To picture this, I have to drop our three-dimensional universe down a dimension to two, so it is a sort of surface, a membrane—"brane" for short—in a three-dimensional "bulk" space. One of the ideas floating around the dockside is that gravity spreads out over all dimensions, while the Standard Model is confined to a lower-dimensional brane. Gravity is then very diluted, and that explains its weakness compared to the other forces.

In some tales of extra dimensions, they are very tiny and curled up. The travelers tell us to imagine waves, vibrations in a string. As far as these waves are concerned, the string is one-dimensional, because the thickness of the string is negligible compared to the wavelength of the waves. But as the waves go higher in energy they go smaller in wavelength, and at some point the wavelength is comparable to the thickness of the string. At that point waves can go around the string, not just along it—a new dimension has opened up. In most stories, this happens so impossibly far east that our fastest ships can never hope to reach it. Though some of the bolder seafarers who join the storytelling assure us that something similar could happen at a high-energy collider, much nearer to home.

Models with additional space-time dimensions were even proposed as a way to avoid the need for a Higgs boson before it was discovered. All the problems with W and Z masses that arise in the Standard Model, and which required the Higgs boson to exist at LHC energies, would have been subsumed into the other changes to physics that happened when we started to be able to resolve the extra dimensions. There would be a need for all the electroweak symmetry breaking stuff, and therefore no Higgs.

One manifestation of the changes to physics that such extra dimensions could cause would be the production of black holes. As we travel east, we are going to higher energies and shorter distances, concentrating more and more energy into a smaller and smaller space. The point of particle colliders, the machines that get us there, is that they concentrate energy like this. A proton in the LHC has a tiny energy compared to a Ping-Pong ball. If all the kinetic energy of an LHC proton were transferred to a Ping-Pong ball, the ball would move about 1 mm every ten seconds. However, there are about 10^{23} protons in a Ping-Pong ball, so that's quite impressive, really, when you think about it.

If enough energy is concentrated in one place, then the gravitational field close to it becomes very strong, eventually even stronger than the Standard Model forces. Strong enough, eventually, to prevent even light from escaping its grip. This is when a black hole is formed; the super-high density of mass and energy warps space and time so much that nothing can escape. The concentration of energy required to do this in a three-dimensional universe is way beyond the capability of the LHC. But in a universe with extra dimensions, this might be possible.[58] Now that we know there really is a Higgs boson, the motivation for this kind of model is somewhat reduced, but there are still old sea-hands who tell the tales, and they can't be completely dismissed. However, we will move on now and listen to another agitated group clustered around a table by the fireplace.

58. There is no chance that black holes produced this way could destroy Earth, by the way; otherwise, this would have happened, and be happening, already and all over the place, because of the very high-energy cosmic-ray collisions that go on around us.

37

Over the Edge

This eccentric group is telling stories of action out east that is still within the Standard Model.

Armed with the newly discovered Higgs boson, the Standard Model can now make predictions for multi-teraelectron volt-scale physics—10^{12} eV and beyond, well above the electroweak symmetry breaking scale. Making these predictions in this qualitatively new regime brings new challenges, though.

As the energy of a collision increases, so do the possibilities for the production of high-energy objects such as high-momentum leptons and photons. Increasing the energy far above the electroweak scale means that objects with masses of around that scale can be produced with high energies, and in high numbers. Several top quarks and W, Z, or Higgs bosons may be produced in the same collision—something we have never seen before. Describing such collisions with any degree of precision requires innovative calculational techniques.

Given that the necessary theoretical advances are made (and they are being made), the experiments can make precise and detailed measurements to test them. The number, reach, and precision will increase as long as the LHC continues to provide data.

And the Standard Model itself may contain some fantastical beasts. An example of one of these is the sphaleron.

Sphalerons are a very exotic prediction of the Standard Model, not something beyond it. And now that we know the Higgs exists, we know they should exist, too. We even know their longitude, roughly. They ought to lie at about 10^{12} eV. This is higher than we can observe in LHC collisions, because even though the proton-proton collision energy is higher than that, the energy of the collision is shared among many quarks and gluons, and so not all of it is available for making new objects. (Remember, colliders are all about concentrating energy). But the energy at which sphalerons should appear is not ridiculously high.

To get an idea of what a sphaleron is, you have to think of the way we describe quantum particles. We use perturbation theory, where we take something that is approximately correct and add small corrections to improve it, piece by piece. We first came across this idea when imaging one, two, or more photons being exchanged between particles. But perturbation theory has its limits. Try this: Grab a soup bowl with a rounded base from behind the bar. Place an olive in the bottom of the bowl.[59] Imagine the empty universe, with no energy, is the olive, sitting at the bottom of the bowl.

Adding a bit of energy corresponds to shaking the bowl slightly from side to side. The olive will roll up the side of the bowl a bit. If the energy is small, it will roll back down again, go up the other side, and oscillate like that. These oscillations correspond to particles in perturbation theory, rippling along through spacetime in their merry quantum way. Adding a bit of energy like this is a small perturbation on the "still olive" scenario, and we can calculate its effect with perturbation theory.

But if you add a lot of energy, the olive can zip over the rim of the bowl and into another bowl that we placed next to it to avoid olive-related chaos. That phenomenon

59. The bar is making a half-hearted attempt to be a gastropub, so it has big green olives in a jar, as well as a bottle of balsamic vinegar and some baskets filled with bread. The regulars are ignoring all of this and sticking to their drinks and stories, so no one will mind if we use them to try to illustrate sphalerons.

of leaping over the hill from one bowl to the next is a sphaleron, sort of. It's a point at which perturbation theory breaks down.

Sphalerons would have been around in the early universe, because the energy density was very high then. And they seem to have played a crucial role in the creation of matter as we know it.

There are various things that don't change in perturbation theory, quantities that are conserved. One thing, in our olive-in-a-bowl universe, is the average position of the olive. It spends as much time on one side of the bowl as on the other; the average is the bottom of the bowl. But if the olive sphalerons its way over the lip into the next bowl, its average position has moved—to the center of the next bowl. The conservation law has been violated. Sphalerons in the early universe violate conservation laws, too, and one that they violate is the number of particles. They can add more quarks and leptons to the universe. This is an essential part of how the stuff we are made of got here at all.

There are other exotica that are sometimes recounted around the less respectable dives on the dock front, most of them beyond the Standard Model, some within it but weird and so far unobserved. We do well to remember in the cold light of day that there is no direct experimental evidence that any of the exotic theories expounded in the dockside pubs is correct. They are all just stories—dreams, or nightmares. But this does not mean that they are completely wild, or completely worthless.

The fact that theories such as supersymmetry, for example, can connect such disparate observations as the rate of neutrinos measured in the Antarctic, to the results of experiments at the LHC, to the observations of galactic rotations shows their usefulness as an aid to exploration. Being able to display data from such different experiments on the same graph, to discuss them in the same context and measure them against a common scale, is important in gauging their relative sensitivity to the unknown, and in hunting for inconsistencies—or, in the case of a discovery, for consistency.

In the absence of observational data, it is terribly hard to distinguish, in the saloon bar of our dockside pub, the ramblings of fantasists from the reasonable guesses of intelligent explorers. Drawing a distinction is important if you are trying to decide how much time and money to put into chasing down the facts behind a story. For a final couple of wild tales from the edges of the map, we return to gravity, and also illustrate one of the ways in which such decisions are made.

38

A Fifth Force

A tenacious dreamer propped up against a pillar by the pool table reminds us that it is also possible that what we think of as fundamental forces just aren't. The quarks, leptons, and bosons of the Standard Model may contain smaller constituents, just as the atoms of the Periodic Table turned out to be made of other things. The solutions to the outstanding problems of the Standard Model might then be found in the interactions of these new, even smaller, pieces. Again, this hypothesis was partly motivated as a way to avoid the need for a Higgs boson. However, there are still versions around in which the Higgs, too, is a composite particle made up of even smaller things.

Such a scenario would involve new fundamental forces and probably mean that the forces we currently think of as fundamental are not, but emerge from some higher-energy, smaller-distance theory that lives out in the far east.

There are also ideas in which the current forces remain, but we add a new one, or change an existing one. Since gravity is a common thread in many of the problems with physics—Dark Matter, Dark Energy, the lack of quantum gravity in the first place—it is very reasonable to expect that general relativity needs to be modified in some way. That's a thought that has occurred to many physicists. However, general relativity is so subtle, and so, well, *general*, that replacing it, or even successfully tweaking it, is a very hard thing to do.

Still, physicists are persistent, and there are new ideas coming forward all the time. One possible tweak is to postulate a new particle that carries a "fifth force" on top of electromagnetism, the weak and strong interactions, and gravity. Maybe out in the far east there is some other form of transport, beyond our road, rail, and airways, and distinct from gravity?

To explain Dark Energy, this force has to affect all matter—as gravity itself does—and operate over large distances. Such forces have been looked for already, and if they affect the motion of the planets in the solar system, for example, they have to be enormously more feeble than gravity; otherwise, we would have seen them already. But if they are enormously more feeble than the gravitational force between stars and galaxies, they won't make any difference to the Dark Energy or Dark Matter problems, so that's a waste of time.

One possible way around this conundrum is a process called "screening," in which the strength of a force depends upon the environment it is in. It is even possible that such a force is screened by matter itself. It is possible to build theories whereby in dense regions of the universe (like Earth, for instance) the force can be hidden, while in empty space the force can operate.

In the case of the Dark Energy problem, which is what some such theories are aiming to solve, this can provide exactly what the data need. The force can make the universe accelerate at large distances while having no measurable effect on the orbits of the planets. As a bonus, this new force can also have a significant impact on the way galaxies rotate, which might at least partially address the Dark Matter issue as well.

The way this new force works is reminiscent of the way the theory of Brout, Englert, and Higgs gives mass to fundamental particles. It involves a scalar boson—a particle like the Higgs boson that has no spin—and it involves the idea of symmetry breaking. If that isn't familiar enough to you to help, consider a pointillist painting, in which an image, and even the color mix within it, is composed of many tiny dots.

When averaged over dense regions of space, a symmetry hides the fifth force. This is like looking at the picture from a meter or so away. The dots are hidden in the colors and landscape of the painting.

Close to the painting, the dots are visible. In the same way, for things the size of atoms or smaller, the averaging doesn't happen, so the fifth force may show up.

And on very long distance scales, space is very empty, so the density is low and again the force reappears. Similarly, a very long way from the painting, it is just a single dot again.

By the standards of wild yarns, this idea seems at least to be testable in an excitingly wide range of experiments. Future observatories will investigate gravity and Dark Energy on astrophysical scales. Precise atomic physics experiments could measure the effect of the fifth force on atoms, and the LHC may also produce, or rule out, some varieties of these weird, so-called chameleon particles.

A theoretical framework, such as is provided by supersymmetry or the Standard Model of particle physics, plays the role of the picture on the box of a jigsaw puzzle.[60] When looking at a jigsaw piece, the picture gives you an idea of where it might fit and how it might connect to the others. Trying to do a jigsaw puzzle without looking at the picture on the box is enormously more difficult than looking at the big picture.

Of course, once you see where a piece might fit in, you still have to try it to see if it really does so. And in the science version of this puzzle, we also have to bear in mind that our picture is almost certainly incomplete (for example in the case of the Standard Model, which definitely describes lots of data) and possibly completely wrong (for example, supersymmetry or extra dimensions, where there are currently no data). But to some extent any picture is better than none, and anyway we don't have much choice. At some point we might fit enough pieces together to realize it was wrong, throw it out, and try a new picture. A sort of paradigm shift that would see a jigsaw-puzzle manufacturer go out of business under the weight of returned Christmas presents.

The framework provided by a good theory, or collection of theories, gives focus to research and makes it harder for a new theory to gain acceptance. A maverick new theory—and there are many around—must either fit with the existing picture or replace it completely. In the latter case, it has to accommodate all the jigsaw

60. In case you were wondering, yes, this is the kind of pub that has jigsaw puzzles on a shelf in the lounge bar, along with backgammon and a Scrabble set with the Q, X, and three vowels missing.

pieces that are already snugly interlocked. This, not a conspiracy of lizards or illuminati (or even hide-bound conservatives), is why a dramatic "Einstein was wrong" type of idea is unlikely to be taken very seriously without a lot of supporting evidence and the ability to accommodate previous evidence that he was pretty much right. It will generally not just postulate monsters out east, it will probably contradict things we already know about lands we have already explored in detail.

In the sense that it emphasizes the importance of the whole and the interdependencies of the parts, the approach described above is a holistic view of science. In an essay entitled "When Scientists Go Astray," the great theoretical physicist Philip W. Anderson describes this as a "seamless web":

> a body of firmly established theory, now extending from physics through molecular biology, which, in many situations, traps dubious observations. Already known laws like conservation of energy, quantum mechanics, relativity, and the laws of genetics, constrain the explanation of any given result in a fashion which can be unique, or nearly so, and makes errors easy to spot. Much of science is "overdetermined" in this sense.

It is by no means infallible, but it is the best we have.

39

Into the Cosmos

"Last call at the bar." Before we leave, a final story—an example of a borderline case, one that is not completely discredited but that most respectable explorers would probably dismiss.

A measurement of the electric charge of antihydrogen was published in 2016 by the ALPHA experiment at CERN. The charge is expected to be zero, and the measurement confirms that, to high precision. That is an important step in the main goal of the experiments, which is to measure for the first time whether antimatter experiences the same gravitational force as matter.

According to the "seamless web" of theory, it should. But the force has never been measured, so from a purely observational point of view, we don't know. It could even be that antimatter experiences the opposite gravitational force to matter, which would mean matter and antimatter repel each other gravitationally—that is, antimatter may experience antigravity.

This brings in another wild tale, the so-called Dirac-Milne universe, in which matter and antimatter do indeed gravitationally repel each other. This would be a very different picture on the jigsaw box, and it is sort of hard to believe that it could describe the observations we already have. These include measurements of particles in colliders, of the distribution of matter in the galaxy, of the radiation left over from the Big Bang, and more. All these observations have driven us to our current picture of physics and cosmology known as the "concordance" model, which describes how the universe evolved after the Big Bang and includes Dark Matter and Dark Energy.

On the other hand, as we have seen, our current picture has substantial gaps.

The Dirac-Milne model claims to address all these problems, and to agree with the data (and the concordance model) on some important features such as the abundance of light elements in the universe and the main feature of the cosmic microwave background. Whether it can be accommodated within the seamless web of current knowledge—and thus profoundly change it—may depend upon whether enough people take it seriously enough to do the calculations and follow through the consequences. Or, as Douglas Adams's fictional detective Dirk Gently might put it, how much time will they take, in a Dirac-Milne model, "detecting and triangulating vectors of the interconnectedness of all things."

The same applies to the other stories of the old sea dogs and young idealists clustered on the eastern shores of Bosonia. One thing is certain—if and when someone really measures antigravitational forces acting on antimatter, or chameleon particles, or sparticles, or anything inside a quark, or the Higgs doing something really odd, or just something that really doesn't fit with the current cover on the puzzle box, someone will definitely have to pick up that story and try.

Meanwhile, as we wander back along the dock toward our hotel one last time, ships are putting out to sea on the tide. The voyages they are making into the eastern oceans will report back and tell us whether or not the ideas and knowledge we have gained out west are a good guide to the east. We have our map of the invisible, featuring the quarks and leptons—fundamental particles according to the Standard Model—deriving their masses from the Brout-Englert-Higgs mechanism and interacting via the electromagnetic, strong, and weak forces by the exchange of photons, gluons, and W and Z bosons to form more complex objects—hadrons, atoms, and eventually the visible world around us. We have good reason to know this is not the whole story. But are any new islands within reach? Or are the land masses of our map isolated? Is there an expanse of empty ocean for great distances eastward? Or are there new islands, continents even, just out of sight and waiting to be discovered? Perhaps one of the ships setting sail right now will answer the question.

Right now, it may be time for a rest, a good night's sleep, to absorb what we have learned and what there might still be to learn. But it is good to know that as we sleep, new ships are setting sail. And they will keep setting sail, as long as enough of us want to know what lies out there, beyond our current horizon.

Further Reading

Hopefully this book whets your appetite for more adventure. If so, then here are a few suggestions:

If you still have shaky sea legs, Chad Orzel's *How to Teach Quantum Physics to Your Dog* will give you much more on waves versus particles and all that. The description of QED in Expedition III owes a debt to a short popular book by one of its discoverers: Richard P. Feynman's *QED: The Strange Theory of Light and Matter* is well worth reading if you want to hear more, from the horse's mouth.

Janna Levin's *Black Hole Blues and Other Songs from Outer Space* is a gripping account of the people (and the persistence) behind the discovery of gravitational waves described in "Gravity: A Distant Diversion." Graham Farmelo's biography of Paul Dirac, *The Strangest Man*, is full of moving and enthralling insights into a great breakthrough and a too-little-known genius. Brian Cox and Jeff Forshaw's *Why Does $E=mc^2$?* is a great way to get to grips with more of the ideas behind relativity, and *Universal* will show you how to begin expeditions of your own, even without a Large Hadron Collider at hand. Sean Carroll's *The Particle at the End of the Universe* gives a good account of particle physics up to and including the discovery of the Higgs, as does Gavin Hesketh's *The Particle Zoo* (and of course my own *Most Wanted Particle* tells you a bit more about the discovery itself). Finally, Lisa Randall's *Dark Matter and the Dinosaurs* has the true flavor of some of the stories we heard in the dockside bars.

Acknowledgments

The idea for the maps started with a talk I gave at the Royal Society, but was developed further in discussions with Tom Avery and Chris Wormell. Thanks also to Tom for patience and suggestions on the text, and Chris for patience with my suggestions on the maps. It was a pleasure working with both of you.

For the US edition, thanks to Nicholas Cizek, Sarah Smith, and the rest of The Experiment team.

The support of Diane Banks and her team continues to be outstanding.

I am privileged to be part of University College London, a university that supports and encourages an amazing diversity of ideas and activities, in research, teaching, and in the writing of books like this. I am grateful to my colleagues there, as well as to my colleagues in the great international community of particle physics.

The love and encouragement of my family is the crust, mantle, and core that supports the entire map of my world.

About the Author

JON BUTTERWORTH is a professor in the Department of Physics and Astronomy at University College London and a member of the ATLAS collaboration at CERN's Large Hadron Collider in Geneva, Switzerland. He writes the *Life and Physics* blog for the *Guardian*, has written articles for a range of publications including the BBC and *New Scientist*, and is also the author of *Most Wanted Particle*, shortlisted for Book of the Year by *Physics World*. He was awarded the Chadwick Medal of the Institute of Physics in 2013 for his pioneering experimental and phenomenological work in high-energy particle physics. For the last fifteen years, he has divided his time between London and Geneva.